The Voice of Reason

A Quarter Century of Fighting Aging

Reason

Selected from *Fight Aging!*

Edited by Kristopher A. Borer

Selected writings from *Fight Aging!* (fightaging.org), 2002–2026.

Editing consists of selection, the removal of quoted third-party material that is not the author's to license, compression, and the transition from blog to book. The arguments, the positions, and the voice are the author's own.

Edited by Kristopher A. Borer.

ISBN 978-1-951974-12-1 (paperback)
ISBN 978-1-951974-11-4 (e-book)
ISBN 978-1-951974-14-5 (audiobook)

Contents

INTRODUCTION

The man who wrote this book is named Reason. Since 2002, first at the Longevity Meme and then at *Fight Aging!*, he has written almost every day about one proposition: that aging is a medical problem — accumulated cellular and molecular damage — and can therefore be treated, slowed, and one day reversed. He wrote it when the idea was a fringe embarrassment, and on as it grew a research program, as the companies formed, as the trials began. More than twenty thousand posts across two decades without a gap — there is no comparable record of a scientific field being argued into existence, in real time, by one of the people doing the arguing.

He came to it sideways, an astrophysicist by training and an early investor in longevity startups before most believed there was a field to invest in. For decades he was an advocate and a chronicler, insisting to anyone who would listen that the work was possible and that someone should go and do it. And then he did it himself: in 2018 he co-founded Repair Biotechnologies — no longer just writing that atherosclerosis should be curable, but trying to cure it. A man argues for twenty years that a thing is possible and that someone ought to build it, and then, quietly, becomes the someone. The book opens on his 2008 statement of that worldview and closes, eighteen years later, on the medicine he is now making. The argument does not change. The world around it does.

The book is drawn from his twenty thousand posts. No one will read the archive, and the best of it is buried under the daily work of commenting on other people's papers — but between those daily notices he wrote essays: on the science, its opponents, its politics, and his own mortality. Those essays are the book.

He made each central argument many times over — that is what advocacy is — so the task has been to find the single best statement of each, and to compress even that. Third-party research he was not free to reprint has been removed; every chapter keeps its date, so the reader can feel the field move beneath the argument.

He is often blunt and often angry — he believes aging kills a hundred thousand people a day, and that those who counsel accepting it are, whatever their intentions, arguing for the catastrophe to continue. He is, in the end, an optimist who has run the numbers and found the problem of aging solvable, the only question how fast — a man who was right early, said so relentlessly, and then rolled up his sleeves.

Reason's writings not only helped me gain breadth and depth in the longevity field, but also understand what approaches would likely be fruitful rather than fateful. I hope that this book can do the same for you, and inspire you to get to work.

— Kristopher A. Borer

July 21, 2026

PART I

Aging Is Damage

The human body is merely a very complex machine, and aging is just another medical condition.

— Fight Aging!, April 2005

Everything else in this book rests on a single claim: that aging is not a mystery but a form of damage, and that damage can be repaired. Here he states it, defends it against the rival theory that aging is a program to be reprogrammed, and lays out what a repair-based approach actually proposes.

The Aging Is Damage Thesis

Fight Aging!, 16 June 2008

There's nothing wrong with becoming old, but everything wrong with aging. Old means experienced, invested, wealthier, time-tested and just all-round better for having been around the block. Aging, on the other hand, is the direct result of biochemical damage you picked up along the way — ongoing deterioration that is a side-effect of being alive. The passage of years brings a constant flow of opportunities for growth and self-improvement, until aging takes away your ability to compete, your ability to take care of yourself, and eventually your life. Someone should look into that.

If you're not one to think much about medical research, you might be under the impression that aging is fairly mysterious, a primal and inevitably metered process quite separate from the diseases of old age. In fact that's not the case. Aging is exactly and precisely the root cause of those diseases of old age, and scientists have a good understanding of what aging actually is, once you get under the hood and start looking at cells and the cellular environment:

- Loss of important cells without replacement over the years

- Accumulation of damaged, malfunctioning and senescent cells
- Free radical damage resulting from damaged mitochondrial DNA
- Accumulation of waste chemicals that cannot be broken down inside cells
- Accumulation of biochemicals gummed together by sugar outside cells
- Formation of amyloid deposits throughout the body
- Breakdown of cell control processes, leading to the runaway growth of cancer

The short story is that aging is damage and change, rust and wear for our biology that is caused by the normal operation of human biochemistry. You can't run machinery without causing wear, and you can't run factories full of machinery without creating waste by-products. Machinery with a lot of rust and wear breaks down in any number of ways, and biological machinery is no exception — just a few classes of wear, rust and buildup of waste lead to a vast array of different malfunctions.

When you can't do anything about the rust, wear and waste, you put on the best face possible under the circumstances and soldier on. Perhaps you convince yourself that the miseries of an increasingly damaged body and mind are for the best. It's a slowly boiling pot, but it's our slowly boiling pot, and it's all we have. We humans are good at that sort of proactive self-deception for the sake of sanity in the face of the inevitable — we've been doing it for a very long time indeed.

All habits outlive their usefulness, however, and self-deception about aging has lingered past its time. These early years of the 21st century are the opening notes in a symphony of biotechnology, an expanding revolution in medicine, research and computation. Already, the laboratories of this decade are far beyond those of the 1990s: cancer researchers are developing targeted, precise cell-destruction therapies applicable to any unwanted, damaging cells; the immune system is being turned to destroy the amyloid buildup of

Alzheimer's disease, and other amyloids of aging will follow; methods to repair and replace damaged mitochondrial DNA are emerging from the laboratory; the regenerative medicine community is deciphering the controls that will make our stem cells do our bidding; and researchers are plundering the bacterial world for enzymes that can break down the damaging biochemicals that build up with age.

If you think aging is inevitable, and that we should make the best of it, then you're probably not helping the world's researchers in their efforts to repair the damage that causes aging. Funding for research is very dependent on the zeitgeist of the age. If most people think that aging is inevitable, conservative funding bodies won't fund research aimed at the repair of the biochemical damage that causes aging. Thus little progress occurs, no one in the public is given any reason to doubt that aging is inevitable, and medicines to repair aging are pushed further into the future, perhaps out of reach for you and me.

Given the choice to be old, wise and better without being aged, frail and ill, wouldn't you choose to repair the damage? It's not a hypothetical question anymore, and the number of years it takes to develop medicines of repair for aging depends upon your answer.

Programmed or Damaged?

Fight Aging!, 12 June 2015

There are many theories of aging, a state of affairs that I would say is really due to past lack of resources put towards building means to treat aging in a targeted way, by addressing specific purported root causes. There has been a great deal of investigation of the biochemistry of aging, and all too little bold experimentation in means to extend healthy life spans. This was in large part cultural for the last generation of researchers, a way to reject any association with the fraud and self-deception of the "anti-aging" marketplace, and thus preserve reputations and the ability to raise funding. At the same time, this meant that greater progress towards longevity might have happened and did not: talk of treating aging became a threat to careers, an unfortunate state of affairs that has only recently abated.

Ways to increase longevity in laboratory species are the tools by which theories of aging can be winnowed and validated. Unfortunately all of the present means of slowing aging are far too general in their operation to serve as good tools in this sense. Take calorie restriction and its alleged mimetic drugs: these approaches change near every measure of metabolism, to the point at which it remains an enormous puzzle to figure out how and why they act to slow aging. What is needed is a new generation of much more targeted therapies — clearance of senescent cells, or clearance of cross-links, or other proposed SENS biotechnologies based on the repair of one single type of tissue damage. Build the treatment, run the experiment, and a lot will be learned from whether or not it extends healthy life. If great progress is made by repairing forms of damage thought to cause aging, that tells us that theories painting aging as a process of damage accumulation are the robust ones.

Ultimately theories of aging matter today because they are used to steer investment in research. At the moment the division of greatest importance is between programmed aging theories and stochastic damage theories. Programmed aging theories would have us think that aging is the direct consequence of a set of evolved changes in gene expression and protein levels and cellular operation, and that these changes cause damage, dysfunction, and death. The right thing to do, if this is true, is to work to alter the operation of metabolism, to bring it back to a more youthful pattern. Stochastic damage theories tell us instead that aging is caused by what is effectively biochemical wear and tear, and that our bodily systems react to the presence of that damage with altered gene expression and cellular operation — but it is the damage that is the root cause. The way forward, if this is the case, is to repair the damage.

Personally, based on my view of the evidence to date, I'm in the latter camp: aging is stochastic damage accumulation. Repairing that damage by following the SENS proposals is very cost-effective; producing full demonstrations of the various treatments in mice is a $1–2 billion, ten-to-twenty-year project at full-scale funding. That's less than the cost of developing a single drug candidate in the Big Pharma world these days. Programmed aging, on the other hand, would direct us into a massive, unending project of trying to fully understand and safely alter swathes of our metabolism. To give some idea of the scale: about a billion dollars has been swallowed up on research of sirtuins in aging over the past decade or so — just a couple of genes out of thousands worth looking at, nowhere near a full understanding of their role, and no meaningful treatments resulted from all of that work.

A spatial evolutionary model published recently gives the programmed-aging camp its best argument — that a built-in life span could evolve because shorter-lived organisms deplete fewer local resources and so leave more for their offspring. It is an elegant idea. But I see programmed aging theories as the road into an endless swamp, a course that might be averted at low cost by making enough progress on SENS rejuvenation treatments in the laboratory to demonstrate their worth in extending healthy life spans — and thereby showing that aging is most likely a process of stochastic damage.

Repair, Don't Tinker

Fight Aging!, 17 December 2013

Recent research demonstrated a five-fold increase in life span in nematode worms. That's actually only half as long as the present record for that species, and both approaches involved tinkering with genes associated with insulin-like signaling, one of the better-studied areas of intersection between metabolism, genetics, and aging. The press picks up on this sort of thing and uses it to wave around wild comparisons with hypothetical 500-year human life spans; wild comparisons in the headline attract attention regardless of merit, and the press is in the attention business, not the truth, sense, and accuracy business. There isn't any merit, of course: one thing that is pretty clear from the data of the past couple of decades is that ways of manipulating metabolism to slow aging only have dramatic outcomes in very short-lived species.

A good example is manipulation of growth hormone metabolism. In mice, the best of these methods extends life by 60–70%. There is an analogous natural mutation in the human growth hormone receptor that leads to Laron dwarfism; those with the condition do not appear to live any longer than the rest of us, though they may be resistant to some age-related disease. That is quite a climb-down in comparison to the results in mice. We see the same trend comparing worms with mice: the best methods of slowing aging produce much greater results in nematodes, which normally live for a fraction of a mouse life span.

This should all make sense if considered from the perspective of evolution. For a short-lived species, surviving a famine to reproduce later requires a great lengthening of life. For a long-lived species, that survival doesn't require any lengthening of life — it requires other types of short-term resistance to privation. So the evolutionary pressures that emerge from seasonal environmental changes are very different for short-lived species, but they are sufficiently ubiquitous across all of evolutionary time to have very deep roots in our ancestry.

The evidence to date, obtained from myriad ways of slowing aging in mice, flies, and worms, suggests that we shouldn't be terribly excited by even a tenfold extension of healthy life through present genetic engineering when it occurs in very short-lived animals. There is no good reason to expect any of these strategies to achieve results of great consequence in humans.

The future of human life extension is very different from this work: it will be based on direct repair of damage rather than altering metabolism to slow the accumulation of damage. Aging is damage, and removing that damage should constitute a reversal of aging. It wouldn't be unreasonable to expect partial repair — partial clearance of cross-links, or replacement of mitochondrial DNA, or removal of some fraction of senescent cells — to have more of an effect on mouse life span than on human life span. But ongoing perfect damage repair should have exactly the same effect in mice and people: the result being agelessness and indefinite healthy life span.

What SENS Actually Proposes

Fight Aging!, 15 April 2014

SENS, the Strategies for Engineered Negligible Senescence, is a detailed research plan for developing the means to prevent and reverse degenerative aging by repairing its causes. It assembles the list of causes from the present scientific consensus on the fundamental differences between old and young tissues — differences not known to be caused by any lower-level process. The potential repair therapies are assembled from the best and latest of research strategies across a range of fields: stem cell therapies, targeted cell destruction, engineered enzymes to break down unwanted biomolecules, immune therapies, and so forth. SENS is shepherded by the SENS Research Foundation but has growing support in the broader scientific community; far from every relevant research program is actually initiated by, funded by, or even known to the Foundation.

Robust mouse rejuvenation has been the first long-term milestone for SENS since its proposal by Aubrey de Grey. No new technology arrives fully formed, and it is understood that the first versions will be flaky, expensive, and much less effective than later refinements. But in the case of rejuvenation this may not matter all that much, as even a somewhat effective form of rejuvenation provides the patient more time in which to wait on those refinements — or even to assist in their development. Robust mouse rejuvenation means a full enough implementation of SENS to double the remaining life expectancy of an elderly mouse, demonstrated and then replicated in rigorous laboratory studies. It doesn't mean an absolutely complete implementation, and it doesn't mean full and absolute rejuvenation: it is a first pass to demonstrate greater benefits than any other approach to date.

When I talk about implementation of SENS in the laboratory, I am almost always talking about robust mouse rejuvenation. I do not mean clinical translation of the result. The path from robust mouse rejuvenation to the clinic might be decades long in the highly regulated US and Europe, but first-generation SENS treatments will hopefully jump into clinics in other parts of the world just as rapidly as did first-generation stem cell therapies. Medical tourism is a wonderful thing, and will probably one day save your life.

SENS is presently divided into seven general categories of damage that cause aging, each largely independent of the others and each requiring a very different approach to repair: a whole different line of research, run in different labs, organized by different groups. Each category is thought to be enough to kill you in roughly a human life span even if all of the others are defeated. The evidence for this is compelling: each of the SENS categories of damage has at least one fatal age-related condition associated with it, for which it is the primary known driver. This is in fact how these forms of damage were first categorized — researchers work backwards from the visible and deadly consequences of aging in search of the mechanisms by which they unfold.

How long remains between now and robust mouse rejuvenation? Ah, there's the rub. How long is a piece of string? It is very hard to predict timelines for research when funding is at a low ebb, even research like SENS wherein it is fairly clear what the researchers should be working on and where the end goals lie. With good funding — say $100 million a year — we could fall back to planning estimates of a decade or so, because with that much money there can be many irons in the fire, and the law of averages begins to smooth out random chance. When there is comparatively little funding, progress is uncertain, and I would be surprised to learn that there was more than about $10 million devoted to directly SENS-related work this year outside the stem cell and cancer research communities.

Robust mouse rejuvenation is not a coordinated single point in time at which all parts of SENS suddenly become available at once. Different areas stand at very different stages of readiness, and some will clearly be done first. The best candidates for early success are, I think, the breaking of glucosepane cross-links and mitochondrial repair of some form. If a method of breaking down the predominant form of cross-links in human tissues is demonstrated to produce benefits, will the world sit around waiting for robust mouse rejuvenation to develop it? Of course not. You may still get nailed by one of the other forms of age-related damage on roughly the same time scale as a normal life span, but it is hard to argue that you will not find improvements to health and function through repair of only one or two forms of damage. Robust mouse rejuvenation is an aspirational goal, but it isn't a dividing line. Results will be piecemeal and staggered, and any result with significant merit will probably be rapidly developed as a treatment in less regulated parts of the world.

PART II

The Damage, One by One

You can plug as many holes as you like, but eventually you're going to run out of fingers.

— Fight Aging!, January 2006

If aging is damage, then the natural question is: what damage, exactly? These chapters walk through the specific kinds — failing mitochondria, cells that will not die, the sugary cross-links that stiffen tissue, cancer, the reprogramming hype — and end on the one he left the keyboard to go and work on himself: the cholesterol that clogs an artery.

The Rogue Power Plants

Fight Aging!, 8 October 2006

Free radicals, and reactive oxygen species (ROS) in particular, play an important part in aging. These are (usually small) molecules lacking an electron needed for stability; they will steal an electron from the first thing they bump into. Like pulling a cog out from clockwork, stealing an electron from a protein or enzyme is usually not good for the finely-tuned biochemical machinery of our cells. The free radical might be rendered safe in the process, but it has left some form of chaos and damage in its wake.

Scientists generally concur that accumulated damage throughout the body due to free radicals is one important root cause of age-related degeneration — but the devil is in the details. The vast, overwhelming majority of those free radicals are generated by your own metabolism as an unavoidable byproduct. The rate of free radical generation increases greatly with age as the basic mechanisms of your metabolism are themselves damaged by the free radicals they created. This is not a one-step process, however. I'll try to walk through it at a high level, cribbing from the mitochondrial free radical theory of aging that was proposed by Aubrey de Grey and working its way into general acceptance.

Within each of your cells are many mitochondria, tiny biochemical power plants that convert chemicals from food to ATP, the basic fuel molecule used by your cells to provide energy for life.

Mitochondria were once a separate organism that came to live in symbiosis with ancestral cells. As such, they brought their own DNA to the party; some of it still remains within our mitochondria, separate from the DNA we carry in chromosomes in the cell nucleus.

Mitochondria have a couple of ways of generating ATP. The more efficient method — oxidative phosphorylation (OXPHOS) — generates free radicals as a byproduct and requires proteins coded in the mitochondrial DNA. It's the predominant way healthy cells make their power, and the free radicals it throws off are so reactive that they rarely travel far: the mitochondrion's own membrane and its own DNA take the damage.

Sufficient free radical damage to mitochondrial DNA shuts down OXPHOS within that mitochondrion, as the necessary proteins can no longer be produced. The mitochondrion switches over to using a less efficient method of producing power, one that doesn't produce free radicals, but has to run at a much higher rate to produce the same level of ATP.

Mitochondria, like most cellular components, are recycled on a regular basis. Components called lysosomes are directed around the cell in response to various signals, engulfing and breaking down damaged or worn components. After the herd has been culled, surviving mitochondria within a cell divide and replicate, much like bacteria, to make up the numbers.

The signal to break down a mitochondrion is triggered by sufficient damage to its membrane: a sign that it's old, leaky, inefficient and needs to be replaced with a shiny new power plant.

But: if a mitochondrion has had its DNA damaged to the point of stopping OXPHOS, it will no longer be producing free radicals that can damage its membrane. So it will never get broken down by a lysosome. When the time comes to divide and replicate, it will replicate its damaged DNA into new mitochondria. None of those new mitochondria will be producing free radicals via OXPHOS, and so will not be recycled either.

One DNA-damaged, non-OXPHOS mitochondrion will eventually take over the entire mitochondrial population of a cell in this way. At that point, the trouble really gets started.

By the time you hit late life, perhaps 1% of your cells are in this state of being taken over by non-OXPHOS mitochondria. As for any neighborhood or city, it only takes a small proportion of dangerous criminals to make life really unpleasant for the rest of us.

Non-OXPHOS mitochondria create a further problem: they deplete NAD+, a carrier molecule cells need for hundreds of processes, because without OXPHOS running there's no way to return the spent NADH it produces back into circulation. The cell's workaround is to export the unwanted electrons through structures on its outer membrane known as the plasma membrane redox system (PMRS) — a process that ramps up specifically in cells taken over by damaged mitochondria.

What do these electrons do? Well, for one, they combine with oxygen molecules — which are abundant in any of our living tissue — to create reactive oxygen species (ROS): more free radicals. So you have the Rube Goldberg system outlined above whereby a few free radicals have caused a cell to become an ongoing, major exporter of free radicals into the surrounding environment. These will make life unpleasant for surrounding cells, but that is most likely not the real problem. ROS just can't travel far enough to explain how a corrupt 1% of our cells can cause a large fraction of the difference between being young and being old.

A more likely target is cholesterol. LDL particles travel everywhere in the body, and when ROS reacts with nearby LDL — and there is always nearby LDL — it produces oxidized cholesterol that goes on to damage biochemical processes throughout the body, arterial disease among them. This, then, seems to be a good candidate for the plausible, systematic method by which a small number of cells can work such varied damage upon your entire body.

Aubrey de Grey has proposed an engineering solution to this problem, based upon this way of looking at it. That is to go straight to the root, and get the OXPHOS process working again by (a) moving mitochondrial DNA into the nucleus, and (b) ensuring that the necessary proteins can make it from the nucleus back into the mitochondria where they are needed.

The reason this isn't as radical as it sounds is that evolution has already done most of the work: of the roughly thousand proteins that make up a mitochondrion, all but thirteen are already built in the cell and imported in from outside, hauled through the mitochondrial membrane by a complex called TIM/TOM. The fix, then, is to make backup copies of those last thirteen genes, modify them so that TIM/TOM will accept them, and place them in the nucleus as well — DNA that is far better protected from mutation than anything left inside the mitochondrion. Get that done and it stops mattering whether the mitochondrion's own copy gets damaged, because the nucleus keeps supplying working proteins regardless.

A great deal of work is needed to make this happen, even with today's biotechnology. But if you don't get started, you'll certainly never finish.

The Cells That Will Not Die

Fight Aging!, 21 June 2006

There are any number of interesting studies and technology demonstrations that extend the life span of individual cells by one means or another. If your cells live longer, you live longer, right?

Well, no, sadly.

Consider this thought: if everyone in your city drove twice as fast, all the extra efficiency would make the local economy grow more rapidly. Right?

But wait — driving twice as fast only really works to save time and improve efficiency under some circumstances. What about residential areas and curvy roads? Can these cars even drive twice as fast all the time without falling apart and causing all sorts of other problems? Doesn't fast driving cause more noise and wear on the roads? Plus which there are always going to be people for whom driving fast is never going to translate into efficiency: they'll be at the beach or in the bar. I'm pretty sure that a large chunk of the economy has no relationship to how fast people are driving, in any case.

Your body is a complex system, and as such its properties — such as health and life span — are not necessarily directly related to the properties of its individual components. Like a city or an economy, your body is a process in motion, with more hidden depths than the mere sum of its parts might suggest. In many cases, tissue and organs have evolved to operate most effectively with a high turnover rate of individual cells, for example. Changing that is somewhat like trying to downgrade a precision engine. Damaged cells need to be removed from the system, and cancer is the ultimate expression of cellular longevity.

There are no doubt subsystems within your body that would benefit greatly from extended cellular longevity; specific tissues in which longer cell life spans would in some way put off the local advance of age-related degeneration. A number of groups are aiming at just this sort of goal in telomere research, attempting to produce therapies for specific, localized age-related conditions. For the most part, however, "extended cellular longevity" seems to mean "lots and lots of cancer."

So, no, a focus on cellular longevity is not the way to be looking at present day research with an eye to applications in healthy life extension. Instead look at cellular damage and its repair: some classes of damage do indeed shorten the life span of individual cells, but that is not the reason why treating this damage is important. After all, some forms of age-related damage lead to cells lingering for too long, cluttering up the joint, or actively proliferating longer than is safe and damaging the body.

This is the way to go: identify the components, identify the damage, prevent or repair the damage. Whether it involves cellular life span along the way is somewhat irrelevant — we care about our healthy life spans.

The Scandal of Glucosepane

Fight Aging!, 8 December 2015

As we age skin and blood vessels lose their elasticity. People care too much about the skin and too little about the blood vessels, but that is always the way of it. Appearance first and substance later, if at all. Yet you can live inside an aged skin; beyond the raised risk of skin cancer its damaged state arguably only makes life less pleasant, and the present state of medical science can ensure that the numerous age-related dermatological dysfunctions can be kept to a state of minor inconvenience. Loss of blood vessel elasticity, on the other hand, will steadily destroy your health and then kill you. Arterial stiffening leads to hypertension and a spiral of cardiovascular remodeling, small blood vessels failing under the strain in ever larger numbers. In the brain, this shows up as the countless tiny strokes behind age-related cognitive decline and, eventually, dementia; more central parts of the system are likely to fail first, by way of stroke, heart attack, or the slower decline of congestive heart failure.

From what is known today, it is reasonable to propose that the two main culprits driving loss of tissue elasticity are sugary cross-links generated as a byproduct of normal metabolism, and growing numbers of senescent cells. Persistent cross-links link proteins in the extracellular matrix together and restrict their normal range of motion; senescent cells, for their part, secrete proteins that break down and remodel the matrix and turn nearby cells hostile.

The most important cross-linking compound in humans is glucosepane. Our biochemistry cannot break down glucosepane cross-links, and as a result it accounts for more than 99% of cross-links in our tissues. This isn't a big secret. Given this you might expect to find researchers working flat out in scores of laboratories to find a viable way to break it down. After all here we have one single target molecule, and any drug candidate capable of clearing even half of existing cross-links would provide a treatment that can both reverse skin aging and vascular aging to a much greater degree than any presently available therapy. The size of the resulting market is every human being, the potential for profit staggering. Yet search on PubMed, and this is all of relevance that you will see published on the topic in the past few years:

- Glucosepane and oxidative markers in skin collagen correlate with intima media thickness and arterial stiffness in long-term type 1 diabetes.
- Glucosepane: a poorly understood advanced glycation end product of growing importance for diabetes and its complications.

This is a tiny output of work. The research and development world is not beating a path here as it should. The thesis is that this lack of enthusiasm exists because the state of tools and processes needed to work with glucosepane has long been somewhere between underdeveloped and nonexistent. No group will choose to work in an area in which they have to build the tools first when there are so many other choices available. This sort of chicken and egg situation exists in numerous places in every field of science and technology, small fields where a great deal might be achieved, but no-one does anything because the short-term rational choice is to do something else in an area where the tooling already exists. This is why we need advocacy and philanthropy, to fix problems of this nature. In recent years the SENS Research Foundation has been funding development of the tools needed for research groups to work with glucosepane in living tissues, and just this year we have seen the first published results: a simple, cheap, efficient method of creating as much glucosepane as needed for ongoing cell and tissue studies. There is now no roadblock standing in the way of any researcher wanting to run up batches of glucosepane, create small sections of engineered skin and blood vessel tissue, generate cross-links in that tissue, and then carry out assessments of drug candidates for clearing those cross-links.

The tools are a big deal, I think. Glucosepane clearance is a very narrow, very small pharmacological problem with a huge pot of gold on the other side. Pharmaceutical companies and established laboratories should be packed with staff running, not walking, to work on this. It is crazy that anyone has to be out there banging the drum to draw attention.

How to Cure Cancer

Fight Aging!, 24 April 2013

Today's topic is the cure for cancer, something of a grail in medicine. It will be challenging to produce, but I think that the difficulty is presently overestimated by much of the public and those in the mainstream of the research community. It is the sheer exuberant variation in cancer — between types, between tissues, between individuals, and even between tumors in an individual — that makes it such a daunting foe. Every cancer is an evolving mess of broken cells with its own character and biochemical quirks.

We stand now in the early stages of a revolution in biotechnology, however, and the rapidly expanding capabilities that brings to the research community are beginning to reveal that, for all their variety, cancers do have at least some shared characteristics and shared vulnerabilities. In that spirit, I offer you three and a half ways to cure cancer, outlined very briefly below:

1) WILT, whole-body interdiction of lengthening of telomeres

WILT is my least favorite cure for cancer, but it is nonetheless hard to argue that it isn't in fact the ultimate cure for cancer. Cancers absolutely depend upon ways to lengthen telomeres, the protective caps at the end of chromosomes that shorten with each cell division, putting a limit on the life span of ordinary cells. A cell with little left of its telomeres stops dividing, destroys itself, or becomes senescent — and thus not much use as a cancer cell. Telomeres are lengthened by the activities of the telomerase enzyme and the mechanisms imaginatively known as alternative lengthening of telomeres (ALT), both of which are abused by all cancers in order to create unfettered growth.

Disable telomerase and the genes for ALT in a human, and the result will most likely be a human who cannot suffer cancer. There is a reason this is my least favorite approach however, and that is that your stem cell populations require the ability to lengthen telomeres in order to continue to maintain your tissues over the long term. A person who underwent a hypothetical WILT treatment would need stem cell transplants or a similar way to refresh all of the different stem cell populations of the body — and there are many — every decade at least. WILT means exchanging the threat of cancer for an arguably greater dependency on medical technology.

Research into WILT is currently funded to a modest degree by the SENS Research Foundation under the OncoSENS program, with a focus on establishing a sufficient understanding of ALT to determine the best and most comprehensive way to disrupt its mechanisms.

2) LIFT/GIFT, leukocyte/granulocyte transfusions

Somewhere out there, someone possesses immune cells (the white blood cells called leukocytes or granulocytes) that can kill your cancer. You might consider this approach to immune therapy as being analogous to first generation stem cell therapies that are presently available in many parts of the world: take someone else's immune cells, grow them in culture, and then transplant large numbers of them into your body, where they work to destroy cancer. This methodology has been shown to produce exceedingly impressive results in mice, such as entire lineages of cancer-resistant mice, but it isn't known why exactly it works so well — which makes it hard to proceed to clinical applications in the US, where a full scientific understanding of the mechanisms involved is generally required.

A couple of small startups are presently working in this area, with little funding and slow progress so far as I'm aware.

This is exactly the sort of application of cell therapies that should do well in the medical tourism arena, and indeed is appearing as an option in some overseas clinics. It is easy enough to implement that any group that can presently carry out stem cell transplants should also be able to manage immune cell transplants. More publicity, signs of progress in obtaining human results, and greater funding for trials would go a long way towards speeding the spread of this therapy and determining whether the results in mice continue to translate well into humans.

3) Targeted cell killing technology, plus the search for commonalities in cancer

Modular targeted cell killing technology platforms with a slot for a sensing system are well in hand in the lab, and are a big part of why the next generation of cancer therapies will be far more effective and far less traumatic than chemotherapy. A great variety of such systems are presently under development: nanoparticles that can be heated by radiation or that carry chemotherapy payloads or RNA interference, engineered viruses and bacteria, trained immune cells, and so forth.

The commonality here is that all of these systems are designed to destroy specific cells with minimal damage to surrounding cells — all that is needed are mechanisms to ensure that these cell-killers only target cancer cells. This largely means discovering suitable markers on a cell surface: specific proteins that differentiate cancer cells by the degree to which they are present, and which are sufficiently general to appear in a sizable population of patients or many types of cancer.

The big uncertainty here is whether or not researchers will find targets shared by many cancers that are sufficiently discriminating to allow enough preferential targeting of cancer cells. It's possible to layer multiple poorly discriminating targets to get a highly discriminating system, however, and there are promising signs of late on this front. You might look at trials involving therapies targeting CD47, for example, which appears on most cancers per the latest research.

If there are enough markers like CD47 out there, then it should be possible to build a comparatively small suite of general cancer therapies that will kill 80% of cancers at any stage, metastatic or not, with minimal side effects. At this point in the development of medicine even twenty different loads for the same basic system looks like a very good thing — and very plausible too, if the process of discovering cancer commonalities keeps going the way it is. All that is needed is one kill mechanism and a delivery platform modular enough to accept the different sensor mechanisms while still being manufacturable at low cost, such as through the use of dendrimers or viruses.

3.5) The mechanisms used by naked or blind mole rats

Naked mole rats don't get cancer, and it appears the same is true of blind mole rats, but for different reasons. Present understanding of the evolved mechanisms by which these animals manage to stay cancer-free for the several decades of their life spans, even while living in an environment that produces a tremendous amount of cellular damage, is advanced enough to have a sensible discussion on how to recreate it in humans.

This really only counts as half a potential cancer cure, however. It does seem to grant cancer immunity, or as near to it as counts, but it is a big question mark as to how hard it will be to safely have our cells start to behave in the same way as those of a mole-rat — even only temporarily. In the case of naked mole rats, the mechanism in question involves the genes p16 and p27, which suggests that it's something that could be accomplished via gene therapy, but much remains to be done in order to find out how much work there is here.

So this is certainly as intrusive a proposal as WILT, i.e. we're talking about altering human metabolism and genetic programming, but far less is known regarding how best to move forward with this strategy. Still, it is probably the case that more researchers are working on it than are in the case of WILT — the cancer community is large and well funded, and the study of mole rats is well recognized these days.

Reprogramming, and What It Isn't

Fight Aging!, 9 December 2021

David Sinclair works on areas of mitochondrial metabolism relevant to aging, sirtuins and NAD, and of late is involved in research into a novel understanding of the relationship between DNA repair and age-related epigenetic change, as well as the use of in vivo reprogramming as a way to reverse age-related epigenetic change. He is first and foremost an excellent self-promoter, however, a job that has come to include authoring books such as Lifespan.

On balance, self-promotion seems a useful trait for people who work on projects of merit. It is a tough job to raise funds and build the necessary networks of allies to push forward the bounds of any field; every advantage helps. It is not such a good trait when the projects themselves are not useful. Self-promoters have a way of cluttering the road to clear debate and understanding, and Sinclair's work is a mixed bag when it comes to utility as the basis for treatments for aging. That said, I think it unlikely that Altos Labs would exist in its present form, devoting a sizable budget — hundreds of millions initially — towards in vivo reprogramming as an approach to the treatment of aging, absent Sinclair spending the last few years aggressively publicizing his work on reprogramming to everyone who would listen. Fortunately, and unlike everything else Sinclair has worked on and relentlessly publicized to date, in vivo reprogramming is an area of research that might turn out to produce a meaningful degree of rejuvenation in old people.

Unfortunately, this still means that one should take everything Sinclair says outside a peer reviewed scientific paper as propaganda, a matter of talking up his position and enabling the companies he is involved in to raise funds and find exits. Self-promotion is a game in which one can win every battle and still lose the war, as people come to filter everything one says through the lens of self-interest and then disregard it. Perhaps Sinclair believes in the position he puts forward in the book Lifespan, perhaps not, but to present epigenetic change as the whole of aging is just too convenient given his portfolio of interests. Maybe he is a true believer in his own work, and that explains it all, but this is also the person who talked up the dead end of sirtuins and resveratrol in exactly the same way fifteen years ago. He overhypes everything that he is earnestly involved in.

I have no horse in this race, beyond a desire to see meaningful progress towards rejuvenation in my lifetime. I've never met the man. I do wish he would temper the way in which he publicizes and promotes his work, however; I think it harms long-term prospects for the field more than it helps.

It is always interesting to see people coming in from outside the field and describing their confusion on encountering the various competing views of aging and possible paths to treatment. The confusion is only exacerbated by the degree to which various parties present narrow views of aging, or present their narrow view of aging as the whole of the picture, or leave out inconvenient points that undermine their position.

One outside reviewer, working through Lifespan for an audience with no stake in the field, arrived independently at much the same conclusion: that Sinclair's confidence in epigenetic change as essentially the whole of aging outruns what the wider research community actually agrees on, and that the cautious consensus view — that aging is probably many separate processes rather than one master switch — has yet to catch up to Sinclair's level of optimism.

The Cholesterol That Kills You

Fight Aging!, 12 August 2020

When people say "cardiovascular disease" in the context of blood cholesterol, they mean atherosclerosis. This is the name given to the build up of fatty deposits that narrow and weaken blood vessels, leading to heart failure and ultimately some form of disabling or fatal rupture — a stroke or heart attack. The primary approach to treatment is the use of lifestyle choices and drugs such as statins to lower cholesterol carried by LDL particles in the blood. Unfortunately, the evidence strongly suggests that this is the wrong approach, in that the benefits are small and unreliable.

Atherosclerosis does occur more readily with very high levels of LDL cholesterol, as illustrated by the early onset of the condition in patients with genetic disorders such as homozygous familial hypercholesterolemia, in which blood cholesterol can be as high as ten times normal. Yet reducing LDL cholesterol levels, even to as much as ten times lower than normal, does very little for patients with established atherosclerotic lesions. One has to look at the mechanisms of the disease in more detail to (a) see why this is the case, and (b) identify which classes of therapy should be attempted instead.

Atherosclerosis is essentially a consequence of the failure of a process called reverse cholesterol transport. When cholesterol becomes stuck in excessive amounts in blood vessel walls, macrophage cells of the innate immune system are called to the site. The macrophages ingest cholesterol and then hand it off to HDL particles. The HDL cholesterol is then carried to the liver to be excreted. This all works just fine in young people. Older people, however, exhibit growing levels of oxidized cholesterols such as the toxic 7-ketocholesterol. Even small amounts of these oxidized cholesterols disrupt macrophage function in ways that are otherwise only achievable through very sizable amounts of cholesterol. The macrophages become inflammatory, cease their work, become loaded down with cholesterol, and die. An atherosclerotic lesion is essentially a self-sustaining macrophage graveyard that will keep pulling in and destroying ever more cells, growing larger as it does so.

The right point of intervention in atherosclerosis is therefore macrophage function. Make macrophages resistant to oxidative cholesterol and cholesterol overload, as Repair Biotechnologies is doing. Or remove oxidized cholesterols from the body, as Underdog Pharmaceuticals is doing. The crucial goal is to allow macrophages to operate normally in the toxic environment of the atherosclerotic lesion; given enough time, it is in principle possible for these cells to dismantle even advanced and sizable lesions. That they do not normally do this is because of oxidized cholesterols or sheer amount of cholesterol, not any other inherent limit.

A large recent analysis of the cholesterol-lowering trial data makes the case bluntly: over three-quarters of the trials show no mortality benefit and nearly half show no cardiovascular benefit at all, and there is no reliable relationship between how much a given trial lowered LDL and how much it reduced risk — some trials with only modest reductions produced benefits while others with much larger reductions did not. That is exactly the pattern you would expect to see if LDL cholesterol was never the right target in the first place.

The Undefeated Monster

Fight Aging!, 2 November 2021

I co-founded Repair Biotechnologies, a company presently focused on developing an approach to rapidly reverse the cholesterol content of atherosclerotic lesions, a goal that is impossible to achieve using the existing panoply of treatments for atherosclerosis. We use gene therapy techniques to provide cells with the ability to safely break down excess cholesterol, enabling the removal of pathological levels of intracellular cholesterol and localized deposits of extracellular cholesterol that characterize conditions such as atherosclerosis (in blood vessel walls) and NASH (in the liver). Atherosclerosis is an important consequence of aging: the structural weakness and narrowing in blood vessels caused by the growth of cholesterol-laden lesions is the cause of death for a quarter of humanity.

Earlier this year, the Foresight Institute invited me to present on the work taking place at Repair Biotechnologies, and the understanding of atherosclerosis that informs it.

I titled the presentation "Atherosclerosis, the As Yet Undefeated Monster" in part as a reaction to a certain type of conversation that recurs when talking to venture capitalists. Many biotech investors, and others too, seem to think that atherosclerosis is a solved problem, and therefore a poor choice of field in which to see a return by supporting the development of new approaches. Many of the best-selling small molecule drugs are statins that have been deployed for decades to treat atherosclerosis by lowering LDL cholesterol in the bloodstream. Physicians reflexively prescribe statins to older individuals. Everyday people obsess about their blood cholesterol levels. New LDL cholesterol lowering drugs that employ modern technologies such as monoclonal antibodies and siRNA are being approved for use on an ongoing basis.

Yet in this environment of decades of ever more attention given to the reduction of LDL cholesterol levels in the bloodstream, atherosclerosis still kills a quarter of humanity. Atherosclerosis is very far removed from being a solved problem. There is an enormous unmet need and ongoing mortality, on a par with the global burden of cancer.

LDL cholesterol reduction as a basis for therapy can lower late life mortality by 20% at most, with many large, robust clinical trials failing to obtain even that degree of benefit. It just isn't the right mechanism if the goal is to produce a cure. It doesn't matter how efficiently a therapy reduces LDL cholesterol in the bloodstream — that 20% mortality reduction appears to be a ceiling. PCSK9 inhibitors do no better than statins when it comes to lowered mortality following control of LDL cholesterol, despite being a much more modern and capable technology.

Atherosclerotic lesions grow over time at a pace influenced by LDL cholesterol levels, but reducing that pace via LDL-lowering therapy cannot reverse established lesions, or even reliably stop them from further growth and eventual rupture. Once a lesion is established, the more important mechanism is the dysfunction of the macrophage cells responsible for clearing cholesterol from blood vessel walls — a mechanism I'll get into in more detail below.

To actually reverse atherosclerotic lesions — to produce a treatment that could legitimately be called a cure — one needs to protect the function of macrophages in the hostile, inflammatory, cholesterol-laden plaque. Make macrophages invulnerable to excess cholesterol and other harms such as oxidized LDL particles, and given enough time they will do their jobs and repair the blood vessel wall.

Here is the presentation itself, tidied up for the page:

So why are macrophages not able to do their job later in life? We need to understand cholesterol transport first. Cholesterol isn't created or destroyed in near all cells, it is rather ingested and excreted. Cells don't break down or get rid of the cholesterol they don't want locally, they hand it off to other cells and parts of the system when they no longer need it. Cholesterol is created in the liver, gets stuck onto LDL particles and goes into the bloodstream, gets stuck in a blood vessel wall, macrophages eat it and then throw it back into the bloodstream to attach to the HDL particles that flow back into the liver. LDL and HDL particles do pretty much the same work when you're young and old, it's the macrophages that stop doing their job. So why exactly do they stop doing their job? Due to a variety of issues — namely systemic inflammation, systemic oxidative stress, and too much cholesterol, although the last is probably not the worst of those three. What this leads to is a feedback loop. Your plaque is a macrophage graveyard, and the signaling of that draws in even more macrophages trying to fix the problem. That is the underlying cause of atherosclerosis.

As you are aware, there's an entire research community and pharmaceutical industry focused only on lowering LDL cholesterol — taking that part of cholesterol transport from the liver to the rest of your body and turning it down. This probably helps a little, since you're reducing oxidized LDL and altered cholesterols, you end up with less altered cholesterol in the plaques, so you're giving macrophages a little bit more breathing room. But it doesn't work to a great enough degree; even if you reduce LDL cholesterol to 10–20% of what is normal in humans, you won't get rid of the plaques, you won't reverse it.

What are the alternatives? Let's start with those that don't work. I mentioned that systemic inflammation is one of the problems leading to atherosclerosis. But if you reduce inflammation systemically, studies suggest you get about the same benefit to mortality as you would get from lowering LDL cholesterol. Which doesn't mean that somebody cannot come up with a way that could do this in a better and more targeted fashion, but the tools available for control of systemic inflammation are really blunt right now.

The second alternative sounds much better, if limited to mice. Reverse cholesterol transport is the pathway wherein a macrophage sucks up cholesterol and hands it off to HDL particles in the bloodstream, which carry it to the liver for excretion. Anything you do in mice to make one or more parts of this system work better, it all works great — up to 50% reversal of plaque lipid content in some cases. But every time it was tried in humans, it failed — there's a whole list of clinical trials over the last 20 years that tried and failed. That tells us that we don't understand something very important about the way in which cholesterol transport is rate-limited in its different steps in humans versus mice.

So our approach is to make macrophages resilient to the environment in old tissues. There have been a number of people trying this, some of it hasn't made it very far, some of it is interesting, and sometimes there is overlap between those two. There is a recent paper with a hypothesis of effect they showed is that if you target lysosomes in macrophages with antioxidants, it prevents the oxidized LDL particles from messing things up, and therefore more macrophages are doing their job — reversing the plaque by 50% in a mouse model. It's entirely possible their hypothesis is wrong and delivering antioxidants is improving something else in the picture, but it's certainly a result that self-experimenters should pay attention to, because these antioxidants are easily available.

Secondly there is the Underdog Pharmaceuticals approach — sequestration of 7-ketocholesterol, which is a highly toxic altered cholesterol, thought to play a big role in atherosclerosis. Lastly there is our approach: engineering macrophages to give the ability to degrade excess cholesterol, whether or not it is altered. The company is named Repair, since I believe that if you're going to address aging and you can't point to something you are actually repairing — a form of damage or dysfunction, where you can clearly say that you are fixing this — then you might not be doing the right thing.

In summary, what we're doing is allowing macrophages to degrade cholesterol and then stepwise approaching the various atherosclerotic conditions in order of number of patients. So starting with an orphan condition — homozygous familial hypercholesterolemia — then you go into larger patient groups as you gain experience doing this. Unlike most therapies, we can actually apply ourselves to any form of atherosclerosis, whether or not it has a genetic cause; we don't care how you got plaques, we just break them down. We've demonstrated AAV delivery of our cholesterol degrading protein — it has a very large effect, 48% reversal of plaque lipids in a month, which is large in the scheme of things as compared to other approaches.

Our goal is to produce a universal macrophage cell therapy. As I said, atherosclerosis is basically the encounter of an aged macrophage with cholesterol, at which point you get a lot of cell death and cholesterol-based plaque. If you overwhelm the existing systems of normal macrophages with excess cholesterol, they can't do anything with it and become pathological foam cells — they don't have a way to deal with that level of cholesterol. So with that picture in mind, the whole spectrum of LDL-lowering drugs really only lowers that input to the problem, and they can't lower it by much, because the macrophage is in the plaque, not the bloodstream. The problem is the plaque that's sitting there, and lowering LDL cholesterol doesn't reverse it.

And the pathological foam cells leading to your plaque bring us to this point, one that has to be made to a lot of people, unfortunately. Your risk of death is not due to LDL cholesterol, it's due to how much plaque you have. It's exactly how much plaque you have and how much high-risk plaque — the soft plaque laden with cholesterol. That determines your mortality. LDL cholesterol, while widely accepted as a surrogate marker, is not the cause of your death. That's why different people can have different levels of cholesterol in their bloodstream and have quite divergent mortality rates.

The point of the exercise is to figure out what we should do differently — and that is making macrophages invulnerable to the plaque-based environment as best as we can. Our idea of "as best we can" is to give a macrophage the capability to break down cholesterol safely in situ. I should say that this is not a trivial thing to do, because a cell is basically an enormous lump of cholesterol — our body uses cholesterol everywhere in the cell membrane. The reason why we never evolved to break down cholesterol when it's harming us is probably because our cells have cholesterol everywhere. So you couldn't evolve something that just chews cholesterol whenever it sees it. And that's why delivering things like the known cyclodextrins that bind to cholesterol is not quite simple either, because the first thing that will happen if you dump a bunch of cyclodextrins into somebody is that their blood turns to mush, because it will consume all your blood cells by hooking all the cholesterol out of cell walls.

So the objective is a safe way of breaking down cholesterol, but only the excess cholesterol, which is what we achieve by putting these specific mechanisms into the cells we're working with. We can demonstrate that by putting these mechanisms into any old cell, and the output is exactly the same — we get a catabolite that is safe and more soluble, and quickly leaves the cell, departing into the bloodstream where it is gotten rid of. What this means is that we can take macrophages and give them the ability to express our cholesterol degrading proteins, and then if you dump cholesterol on those cells they remain competent and able to ingest cholesterol and dispose of it — that's what you want in your plaque.

So going forward, we take induced pluripotent stem cells (iPSCs) from mice or humans and disrupt the lines to make them universal, stripping the surface markers that would otherwise make them recognizable to a host immune system. We then differentiate those iPSCs into macrophages that express our cholesterol-degrading proteins, which is how we produce a cost-effective cell therapy. We've injected mice with the first of these cells over the last month or so and should have initial data by the end of the year.

And then what we do with this is a stepwise approach: through the orphan indication of homozygous familial hypercholesterolemia, with very few patients and a much easier FDA process, then to the heterozygous familial hypercholesterolemia indication with more patients, and then to the large high-risk subpopulations of atherosclerosis, possibly tens of millions of patients at the end of the day. These are all people who will have medical imaging carried out to show the presence of high-risk, cholesterol-laden plaques. Ultimately we think you can take the lion's share of death — of that 27% of deaths caused by atherosclerotic diseases — and use technology such as ours to remove that cause of death from the human condition. How long is it going to take? Who knows, but the most high-risk population is where we start.

Clearing the Plaque

Fight Aging!, 15 August 2024

The Forever Healthy Foundation is the non-profit arm of Michael Greve's efforts to advance the state of the art in the treatment of aging as a medical condition, with a particular focus on the SENS damage-repair philosophy first developed by Aubrey de Grey and collaborators more than twenty years ago, and still influential today.

The for-profit side of Michael Greve's activities is represented by Kizoo Technology Ventures, an investment fund that focuses on the small number of biotech startups that are developing therapies that seem plausibly capable of producing actual rejuvenation — actually repairing one or more forms of molecular damage relevant to the SENS view of aging.

The two sides meet at the Rejuvenation Startup Summit events, where companies present on their ongoing work, hosted by the Forever Healthy Foundation staff. This year the Summit was held in May in Berlin, with the usual selection of high quality projects presented. It seems clear that the most important therapies in the treatment of aging will be those that can reverse aging by repairing the damage that causes it. Everything else is a sideshow, small gains, a lot of light and noise but little action. The Forever Healthy Foundation has been releasing videos from the 2024 Summit.

What follows is the tidied-up transcript of my own presentation, as CEO of Repair Biotechnologies. There were a great many other interesting technologies on exhibit at the Summit: Deciduous Therapeutics, with a promising one-time senolytic modulation of the immune system; Mogling Bio, developing improved versions of CASIN, a treatment that improves stem cell function in a lasting way following one round of therapy; Revel Pharmaceuticals, building ways to clear advanced glycation end-products from tissues; Nanotics, with a clever approach to very selective, rapid, effective clearance of problem signal molecules from the bloodstream; Cellvie, working on the logistics needed to make mitochondrial replacement a viable and widespread option for rejuvenation; Intervene Immune, conducting human clinical trials of a growth hormone based approach to regrow lost thymus tissue; and Rubedo Life Sciences, working on prodrug approaches to senolytics to clear senescent cells.

As it says on the first slide here, we have achieved significant reductions in the burden of atherosclerotic plaque in mice, and we intend to continue the way we started. With respect to the current standard of care for atherosclerosis, the big problem here, the thing that I'm going to beat you over the head with for five minutes or so, is that present treatments cannot greatly, rapidly, or reliably reduce the existing burden of plaque. But we can, and that is our value proposition.

Atherosclerosis is probably going to kill you if nobody does something about it. The rupture of fatty plaque that grows in blood vessels kills 27% of us directly via heart attack or stroke — and that is in a world in which everybody who can use statins is using statins. Further, that doesn't account for the people who are killed less directly by the reduced blood flow to the heart that leads to heart failure. In short, atherosclerosis is not a good thing, and is the greatest single cause of human mortality, greater even than cancer. The more plaque you have, the bigger the problem: looking at great data from a 2004 Dutch study, your mortality risk is five times greater if you have five or six plaques evident in imaging than somebody who doesn't have those plaques. The mortality risk is a very large range, by the way, as the type of plaque matters greatly — how much fat is in it, how much cholesterol is stuck in that plaque, how likely is it to rupture. Either way, if your cardiologist can see five plaques in your major arteries you are not in a good place. You want to do something about it, but today what can you do about it? Not an awful lot.

Let me call your attention to two meta-analysis studies as good examples — there are any number of others we could look at — in which researchers looked over several dozen studies and thousands of patients. The percent atheroma volume is the percentage of the volume of the artery that's obstructed by the plaque. We are looking for percentage change in that number. You want it lower, but if after 18 months of statin therapy the outcome is a range of minus 5% to plus 3%, well, that is not a great form of treatment. The standard mean difference is even worse, as it is essentially zero. So if you take statins for 18 months, your existing plaque is still going to be there at the end of the day. But the plaque is the story! It is the most important part of this, because there are large studies that show that if you can produce a 1% reduction in your percent atheroma volume — just 1% — then the reduction in cardiovascular event risk is much bigger, on the order of 20%. But that takes 18 months or thereabouts, and only a subset of patients actually achieve that outcome no matter how low their blood cholesterol. There are people taking combinations of the modern PCSK9 inhibitors and statins who, despite very low LDL cholesterol levels, have existing plaque, and it's going to stay there.

Why am I pointing out this 1% figure? That leads us to the next point: in mice we can reduce plaque cross-sectional area in the aortic root by 17% over only six weeks of treatment. We look at a cross-sectional area at random in the body of the aortic root plaque to get that 17% figure, and if one assumes it will hold all the way through the slices of the plaque, then this is a 17% volume reduction, not just cross-sectional area. This is a big deal. The importance here is that we can reliably achieve this in mice, and we hope to be able to do the same in humans. You might ask how we achieve this outcome in mice, and the short answer is that we use an LNP-mRNA gene therapy to make cells clear out localized deposits of excess free cholesterol. Excess free cholesterol is a feature of aging, and unfortunately it is also a feature of obesity — which is one of the reasons why a lot of the consequences of obesity appear similar to those of aging. Cholesterol is transported around the body; it isn't really made or destroyed locally, and that complex transport system breaks down in ways that give rise to localized excesses of cholesterol. That localized excess overwhelms the cell's ability to make free cholesterol safe by esterifying it, or incorporating it into cell membranes, or attaching it to transport particles, and that free cholesterol is toxic. It is toxic in the liver particularly because the liver is the center of cholesterol metabolism; your liver function is greatly diminished and harmed by this excess free cholesterol.

Now unfortunately this free cholesterol is undruggable. There is no "break down the cholesterol" mechanism operating throughout the human body, or even in liver cells, that you can adjust with a small molecule. You can't bind and sequester enough free cholesterol with something like a cyclodextrin to get rid of these excesses without killing the patient first, by removing too much cholesterol from cell membranes. In particular, LDL cholesterol in the bloodstream, which is the target of lipid lowering therapies such as statins, has a tenuous relationship at best with a localized excess of cholesterol that might occur at some place in your body. You can lower LDL cholesterol in the bloodstream as much as you like and it won't really do that much to a localized excess of cholesterol.

That said, there are human proteins that can act in conjunction to degrade excess free cholesterol. These proteins are just not expressed together in near all cells in our body. We took the best of these proteins, turned them into an optimized fusion protein, encoded that fusion protein in messenger RNA, and put the messenger RNA into a lipid nanoparticle that is targeted to the liver. We introduce the therapy via intravenous injection; it travels to the liver, it expresses our fusion protein in liver cells, and that clears out the excess free cholesterol in the liver. The liver is thereby restored to homeostasis, and the result is systemic benefits throughout the body via removal of this age-related and obesity-related contribution to disease and dysfunction. It is not just atherosclerosis that is affected, but many other conditions as well.

We are heading in the direction of our first human trials, and our focus is on the rare genetic disease of homozygous familial hypercholesterolemia — HoFH, as I'll call it from now on, for the obvious reason that it is hard to say repeatedly. HoFH patients have enormous blood cholesterol levels, and over a lifetime that means they exhibit accelerated atherosclerosis. Absent any sort of intervention these people die in their 30s — they are not in good shape. There are very few HoFH patients, and this rare disease status means the barriers are lower for FDA approval, and potentially fast track approval. By late 2025 we are going to have efficacy data in a high fat diet model in cynomolgus macaques, running in parallel with getting ready for the first clinical trial for HoFH in early 2026. Of course questions of time are essentially questions of money — all of this depends on raising a Series A round this year, enabling licensing, GMP manufacture with a CDMO, IND-enabling studies, and all the rest of it. Everything is lined up, it just needs the funding and off we go. This timeline is 18 months to the IND submission, and you might feel that this is a long time — but this is in fact very fast in the world of biotech, as I'm sure many of you are aware.

But coming back to the non-human primate studies, here is why I wanted to mention our work there. There's a company called Verve Therapeutics — you might have heard of them, because they recently ran into trouble with one of their clinical trials, but back in 2021 they were riding high, having just gone public at a valuation approaching a billion dollars. At that time, the only data they had was in non-human primates; they had not conducted human trials at that point. Our therapy is objectively better than theirs, which is just a better statin, in brief: it lowers LDL cholesterol, and thus cannot possibly reverse plaque, just as is the case for all of the other LDL lowering technologies. As I said, we will be in this position about 15 months from now with a much better therapy, which is hopefully food for thought for those who might be thinking about investing in us.

I should say that while atherosclerosis is why we are running this program, the starting point, it is not where we will stop. This is a first-in-class therapy that's going to spawn half an industry's worth of further effort, research groups and companies tackling the many age-related conditions that are aggravated by free cholesterol pathology. This really should be a hallmark of aging, and I'm sure that as the hallmarks expand somebody will add it. We have demonstrated that we can reverse the liver fibrosis of metabolic dysfunction-associated steatohepatitis (MASH). We have also demonstrated over the course of that preclinical work that this form of therapy is likely relevant to type 2 diabetes as well, having shown improved glucose tolerance and insulin levels following clearance of excess free cholesterol from the liver. Further, there are numerous neurodegenerative conditions in which the lipid metabolism of the brain is relevant — patients and models exhibit cells with lipid droplets in the brain, there is clearly something going wrong in there, and we believe excess cholesterol is probably relevant. Similarly I can point to a range of evidence in the cancer field, in immunology, and in a number of other rare diseases, in which cholesterol metabolism runs off the rails in some part of the body, and where a therapy like ours could help. This work will obviously be undertaken by an industry that comes after us, not by us — atherosclerosis is a very big problem in and of itself, but there is an even bigger pipeline here.

In the remaining ten minutes I'm going to take you through a very brief tour of our results in atherosclerotic mice. This is the high level only — we have much more data, but the most important data is of course what happens to the plaque. We'll start with serum samples from a time series study. These LDLR-knockout mice are the model for HoFH; the human patients also have loss of function in the LDLR gene, meaning the liver can't take up cholesterol, so the blood is filled with gunk: lots of cholesterol, lots of triglycerides. When you pull serum from these mice it is actually opaque, unlike normal serum, which is clear. In this time series study we injected the mice once with our therapy, and then sacrificed them at various time points afterwards. By 96 hours, a single injection of our therapy has essentially reversed the problem in the serum. The serum is back to being clear; we've removed the gunk, and that contributes to an impressive effect on the plaque.

Next up is the data showing the 17% aortic root plaque reversal that I mentioned earlier. To fill in the details, we took LDLR knockout mice and placed them on a high fat diet for 16 weeks to make them severely atherosclerotic, then gave them six weeks of weekly injections with our LNP-mRNA therapy. We used a very broad range of doses, from 0.25 mg/kg to 1.50 mg/kg. Those who know the mRNA space will know you won't go far wrong by picking something between 0.5 mg/kg and 1.0 mg/kg — many therapies seem to resolve to that dose in the end, and perhaps there is some universal reason for that. In any case, we picked a range of doses and they all worked to reduce plaque, and when averaging across the groups we see the 17% reduction in plaque cross-sectional area. There is nothing stopping us from extending these study timelines — waiting a few weeks and introducing another six weeks of therapy with these mice. We haven't done that yet, but in principle we should observe a greater regression of plaque.

Looking at further data, we took the mice and put them on a treadmill at study close, prior to sacrifice. Bear in mind that these are not mice that will win any prizes for exercise capacity, because they are fat and sedentary, but all of the treated mice, at whatever dose, regained cardiovascular function to a very sizable degree versus the mice who were still impeded by the severe atherosclerosis they were suffering. I want you to note that this is a very broad therapeutic window, with no side effects, and yet benefits were observed at all doses tried.

Now we'll move over to the APOE knockout mice, another accepted model for atherosclerosis in the general population. We conducted a study in which we compared our therapy with a statin. It starts the same way as before: put the mice on a lengthy high fat diet and then provide six weeks of treatment. In this case it was a single LNP-mRNA dose at 1.0 mg/kg alongside a reasonable 5.0 mg/kg daily dose of atorvastatin, a dosing level that will not provoke liver problems in the mice but should still provide a visible effect size on lipid metabolism. As the data shows, our therapy is not only much better than statins, but also synergizes with statins. All other treatments are complementary to ours, be that the Cyclarity approach, or statins, or PCSK9 inhibitors — they can all be used in conjunction. It is an enormous industry; there is room for everybody. Again, this six week treatment could be repeated — you could take a break and conduct the treatment again for six weeks, and we would expect larger results on plaque composition and size.

An important point is that we're not just removing lipids from the plaques and thereby stabilizing them, but we're also increasing the collagen content of the plaques. The lipids are effectively being replaced by collagen, and the result is a more fibrotic plaque, which is stable and safer. Further, we're not replacing lipids with calcium — the plaque is not becoming calcified. I'm not going to present data on that topic, but I will say that in these models we do see a reversal of calcification in these plaques. You should probably take that with a grain of salt, given that the LDLR and APOE knockout models are not specifically models of calcification — if you put 100 of these mice on a lengthy high fat diet, you will maybe see 10 or 20 that exhibit a large degree of calcification in their plaques, not really enough to robustly draw conclusions.

I'll finish up by mentioning our team. I think most of you know Bill Cherman and myself, the co-founders; we come from a mixed investment and patient advocacy background in the longevity community. Morad Topors is a tremendously talented researcher who is here today, so you should take the chance to talk to him — he has a very good background in cardiometabolic disease and is probably the world's leading expert at this point on clearing cholesterol for therapeutic effect. Bobby Khan, our CMO, is a very reputable, well-known cardiovascular physician who has put drugs through the FDA and gives us good advice on our forthcoming clinical program. Of course I really should mention that none of this would be possible without our very talented lab team, who in this picture are standing outside one of the two restaurants in Syracuse, New York that you have to go to if you go to Syracuse, New York. Now you know — if you're there, go to this restaurant. Collectively this is now the world brain trust on clearing cholesterol for therapeutic effect, in that this program is only being conducted by us: nobody else has this, nobody else owns this, nobody else is working on this. We are the first-in-class approach to reversing cardiovascular disease.

PART III

Why Nobody Listens

It is a strange world that we live in, in which we have to argue — actually debate with people who earnestly hold the opposing view — that more of us living for longer, in better health than is the case today, is a good outcome.

— Fight Aging!, May 2021

The science is only half the fight. The other half is that a great many people do not want longer lives, or believe they should not want them, or have built careers on explaining why death is a gift. Here he takes on the opposition — the bioethicists, the columnists, the comfortable fatalism of the culture — with less patience than charity.

Why the Public Shrugs

Fight Aging!, 26 May 2014

Most people don't really care as to how many years of life they have left. It isn't an interesting topic for them, and is thought about rarely if at all. If pushed for preferences, the average fellow reverts to not wanting to rock the boat, to go with the observed defaults: to live as long as his grandparents, or just a little bit more than his peers, enough to make the point without being crass. But this is not really an expression of preference, it is simply going with the flow, the knee-jerk desire for conformity and hierarchy. Of course most people are in relatively good health and a demise by aging is still decades away, which might as well be never given the human psychology of time preference. Life is busy, and it is easy to push future concerns to one side in favor of the day's distractions, needs, and pleasures.

You can conduct a survey of such views yourself: talk to healthy people, and you'll find that most are surprisingly indifferent to the length of time they have left before aging greatly harms and then kills them. It only becomes a pressing concern when that time remaining drops into territory that instinctively makes a person uncomfortable: a month or a year, perhaps.

One of the aspects of ancient Stoic philosophy is that length of life and even present health is of little concern when it comes to happiness. Will triumphs over circumstance, as present state of mind is under the control of anyone who strives for that goal. To quote Epictetus, the Stoic can be "sick and yet happy, in peril and yet happy, dying and yet happy, in exile and happy, in disgrace and happy." In times when life was far more perilous and fraught with sickness and discomfort than is the case now, these sorts of considerations were not academic exercises.

The Stoic and Epicurean view is not just historical: it holds that a life, once happy, does not become any happier by lasting longer — that we don't need long or indefinite lives in order to be truly happy and content.

There is far more to Stoicism than this small slice relating to happiness and length of life. It certainly isn't the "philosophy of what I'd do anyway," but this small slice does more or less reflect the default view of the public at large.

But so what if we have a world of people who hold Stoic views with respect to longevity and years of health remaining, but without any real intent of doing so. Why does this matter? It matters because defeating aging and age-related disease is a grand goal in medicine. Even if the prototype technologies might be pushed to the point of demonstration in mice for a billion dollars and ten to twenty years of work — well within the purchasing power of a large multinational research company or collaboration of billionaire philanthropists — vast resources and many hands will be needed to translate that into a full, mature, worldwide clinical industry. At this sort of scale, at least a sizable minority of the population must support the goal in question for it to have a hope of moving from possibility to reality. There are many grand engineering projects and industries that might already exist in a different world but which in ours have too little public support to move forward rapidly: irrigating the Sahara, a low-cost orbital lift industry, commercial small-scale nuclear reactors, and so forth.

The near future of human longevity is not just a matter of research and building the tools needed to repair the damage of aging, but it is also, vitally, a process of convincing enough people that this is even worth doing. Anyone who has spent time looking at exactly what it means to be old, at the drawn-out pain and suffering inherent in the age-related failure of all organs and bodily systems, might be forgiven for thinking that we live in a madhouse. But nonetheless, most people simply don't care, and won't, until such time as they are in the clinical system asking how their pain and lost function can be assuaged — which is far too late.

Deathism

Fight Aging!, 4 March 2007

Sherwin Nuland is the author of a pro-death piece on the work and aims of biomedical gerontologist Aubrey de Grey, published to much furor from the healthy life extension community two years ago. Nuland stands in much the same camp as another noted deathist, Leon Kass — both labor under the singularly strange viewpoint that the present length of human life is just as it should be, and we should do nothing much to change it.

In his book The Art of Aging, Nuland lays out his belief that eternal life is a bad idea, dressed up in a litany of pragmatic, scientific, demographic, economic, political, social, emotional, and secularly spiritual justifications for why we are better off dying when our biology decrees it. Not nice, in other words, to fool with Mother Nature.

This position is nonsensical. Let us start by asking whether a Nuland transposed to 1900 would argue the same point, back when life expectancy at birth was 40-something years and all older people suffered far more than they do today. Think about that a moment: in an upward trend over centuries, there have always been those who stood astride the engine of progress and said "enough — it's fine just here, danger lies ahead." The opponents of progress have always been wrong to do so, and they're just as wrong today.

Our present length of healthy life and life expectancies at various ages are already highly unnatural; clearly Nuland does not advocate a return to a primitive era of no medical intervention and short lives of suffering, yet there is no such thing as "dying when our inherent biology decrees that we do" today. To support such a thing is senseless. From cradle to old age, modern medical technology increases our health and our odds in the life expectancy game — and those odds continue to improve.

As for Kass, I see this viewpoint as an unthought conservatism (in the sadly fading, non-political sense of the word) — the knee jerks in response to change, and the knee's owner thereby justifies the present state of being as the best possible. The gut drives the brain. But you don't have to look far afield at all to see that such a position is utter nonsense, the "justifications" all smoke and mirrors. Worse, it is a veil, an excuse of the flimsiest sort, for failing to work to prevent billions of unnecessary deaths from age-related disease in the decades ahead.

Remember this: there are many people in the world who want you and everyone you know to suffer and die on their schedule, far sooner than you might. There are people in the world who would suppress all medical research for increased longevity — exactly the sort of research that has increased the healthy life expectancy of the old over the past fifty years, and will accelerate this trend going forward. If everyday folk like you and me go along with these deathists in silence, if we do not loudly point out the nature of the Emperor's clothes, then we will get what we deserve — suffering and death for failing to stand up for ourselves, failing to support longevity research, and failing to build a better future for all.

Death Does Not Give Life Meaning

Fight Aging!, 16 August 2012

Medicine has always been driven by the urge to remain alive, and deliberately tackling the root causes of aging — rather than addressing named diseases one by one — is simply the next logical step in that process. We know more than enough to get started down this path, and a few organizations are working on it even today, though far from enough, and with far from enough funding.

Consider a world with the means to prevent aging — say, through a package of therapies that a person undergoes every twenty years or so, repairing the cell and tissue damage that causes age-related disease. These therapies prevent and reverse the build-up of damage, allowing a body to continue in good health indefinitely. There is no good reason for them to be any more expensive than the average run of clinical treatments today: they would require little time from a physician, and would operate in much the same way for everyone, allowing economies of scale in production and distribution.

In such a society, all of the pressures associated with the short span of life we presently enjoy evaporate. We are so steeped in that omnipresent pressure of time that it's somewhat hard to envisage what a society without it would look like. Every strategic decision that we make in the course of our lives is based on time — that we have ever less of it remaining, the clock is ticking, and we have only a few shots at getting anything significant accomplished. It requires a decade to become truly talented in any particular profession or skill, for example, and at least a few years to figure out whether or not we can follow through to that level. That is a vast investment of time when we only have a few decades in which we are at our prime. The same goes for careers and relationships of any significance. We are pressured and choices have great weight precisely because we must forever give up an ocean of possibilities in order to swim in any particular pool.

There is a related school of thought among those opposed to engineering longevity: they say that the pressures of time created by the fact that we age to death due to our inadequate medical technology are a good thing. To me this has the look of rushing to justify what is, regardless of what might be, but they argue that the industry of individuals and humanity as a whole requires the deadline of dying; that without it, no one would accomplish anything. They look upon the unending holocaust of death and destruction caused by aging — 100,000 lives every day, all they knew, all they could accomplish in the future, all they might have done, erased — and say it is necessary.

This is a hideous nonsense, serving to illustrate that little but a veneer separates us from the barbarians who actively slaughtered millions in past decades. It is true that rapid progress is very necessary in today's world — but we need it because we are dying, and the only way to save ourselves is through technological progress. The faster the better, every increment of speed representing countless lives that might be saved on some future date. If more people were more aware and more interested in doing something about this, we might move faster yet towards the biotechnologies of rejuvenation. Unfortunately, for all that each and every human life is shaped completely by the foreknowledge of future disability and death, all too few are willing to help change this state of affairs.

But so what if the medical technologies that can prevent death by aging make our societies slower-paced, more considered, less energetic? I'm not of the mind that this is a terrible thing — free-wheeling use of a resource is characteristic of wealth, and when we are wealthy in time, we will have the luxury to use it in ways that presently make little sense, or are called wasteful. Caring about waste is a sign of poverty, a sign that we don't have enough of whatever we worry about wasting, which in turn suggests we should do all we can to accumulate more of it. Besides, I don't for one moment believe that the slowing of economic engines and technological progress will in fact happen as feared by those who advocate for the continuation of mass death and suffering. There are all sorts of economic pressures upon human action that have next to nothing to do with aging and our current all-too-short span of life: consider the shifting desires for security, food, property, knowledge, and novelty, for example. The timescales on which those urges operate will not much change in an ageless society, as people will still have the same human nature as exists today. There will continue to be dynamic and ever-changing industries devoted to keeping people fed, clothed, and entertained.

These responses to irrational fears are, at the bottom line, unnecessary. Tens of millions die every year, and hundreds of millions more suffer terribly on their way to that end. There is no argument that can possibly outweigh the need to address what is by far the greatest cause of death, suffering, and loss in the world — yet, for some strange combination of reasons, many people keep trying to find one.

The Cost of Living Longer

Fight Aging!, 2 December 2013

A fair number of people stand opposed to efforts to extend human life. A wide variety of reasons are given, including variants of the mistaken belief that life extension must mean some form of maintaining old people in increasing infirmity and pain for longer. In other words, that improvements in human longevity will look just like those obtained from the past few decades of progress in medicine: expensive and ultimately futile attempts to patch over age-related conditions without doing much — or anything — to address their root causes. Daniel Callahan, a long-standing opponent of efforts to extend longevity, apparently holds this view.

Callahan, at 83 and himself kept alive by a seven-hour emergency heart operation and nightly oxygen for emphysema, argues that modern medicine is very good at keeping the elderly expensively alive on borrowed time — that a small fraction of the population, mainly the old, consumes the great bulk of health care spending on chronic illness and end-of-life costs, and that the longer lives medicine has given us have run in exact parallel with the explosion in chronic illness and its costs. Can we possibly afford, he asks, to live even longer?

This is not exactly correct. From an individual perspective the outcome of modern medicine is a longer life spent with less chronic illness. A rise in counts of age-related dysfunction is the result of many more people living longer — and this is a good thing. Those people enjoyed more years of active, healthy life before becoming sick. But the opening premise is true: modern medicine is indeed good at keeping people alive when they would have died in past decades or centuries, and maintaining a heavily damaged system of any sort is an expensive proposition.

But that is today, not tomorrow. The root of Callahan's erroneous view of the future of medicine lies in thinking that future attempts to extend life, deliberately and by targeting the root causes of aging, will in any way be the same as past medical development. In the past, no efforts were made to address the underlying causes of age-related degeneration, and all therapies ameliorated or patched over damage after it occurred, with treatment usually taking place in the very late stages of dysfunction. As an approach this is doomed to be both a failure and expensive. Yes, it provides considerable benefits in comparison to doing nothing, but it is a world removed from actually going after and removing the root causes of aging and age-related disease. Comparing the shift to the importance of the advent of germ theory is not an exaggeration.

Callahan looks at what is and projects it into the future: the same fundamental failure that bedevils most prognosticators. The scientists have been saying for years that the next generation of medicine to treat aging will be fundamentally different from what has come before, but all too many people are not listening.

Wesley J. Smith is another writer opposed to life extension — and indeed to pretty much all transhumanist projects aimed at overcoming the present limitations of the human condition. He critiqued Callahan's piece, pointing out a little of the double standard inherent when someone who is taking full advantage of modern medicine encourages others to die on a shorter time frame. It is probably the only thing Smith will write any time soon where I actually agree with some of what he is saying. This seemed a rare enough event to point out.

Smith zeroed in on the double standard: Callahan, having availed himself of an expensive heart operation to survive to 83, is in no position to lecture the rest of us about accepting death on schedule. And when Callahan suggests that public money for radical life-extension research is unwise, and that the elderly ought somehow to be prevented from living longer, Smith presses him on what exactly that means — some kind of moral duty to die, or, in less genteel hands than Callahan's, something more insidious: refusing efficacious medical care to the old that the young would still receive.

There are factions who are quite willing to be up-front about plans to limit medical care for the elderly, though this is usually a dialog that takes place in the context of the huge mess that is government-funded provision of medicine. The inevitable results of that mess are rationing, poor service, research and development grinding to a halt as the incentives to improve vanish, and, ultimately, running out of other people's money to the point at which you see a collapse of the sort that ended the Soviet Union, or which is ongoing now in Venezuela.

Even aside from this, however, there are those who oppose even free and open efforts to extend healthy human life. Environmentalist movements are perhaps the most prevalent of these. All in all there are few topics beyond life extension that are so able to bring otherwise sensible, ordinary people to espouse hideous philosophies, such as demanding that the current toll of suffering and death — 100,000 lives lost to aging every day, while hundreds of millions suffer age-related pain and dysfunction — continue without any attempt made to improve the situation.

Is Wanting to Live Selfish?

Fight Aging!, 11 February 2015

In a recent Reddit discussion, philanthropist Bill Gates dismissed the funding of longevity research as egocentric — with malaria and tuberculosis still killing the poor, he suggested, it seemed self-indulgent for rich people to spend their money chasing a longer life for themselves, though he added that it would be nice to live longer.

The comments were of course replicated far and wide in the echo chamber of the press: Gates has a soapbox of an enviable size, even among billionaires. The context here is the Silicon Valley network of wealth that, in quite different ways, funds both Google Ventures' new California Life Company investment and the less flashy but so far much more important rejuvenation research and advocacy organized by the SENS Research Foundation. The philanthropy of the Bill and Melinda Gates Foundation by comparison is focused on improving the lot of the poor directly, as a matter of delivering and implementing existing technological capabilities rather than building completely new things — though of course they do fund malaria and other infectious disease research in a big way, which is definitely cutting edge biotechnology.

Both advancing the cutting edge and delivering existing capabilities to the poor are viable approaches to making the world a better place. There must be progress at the cutting edge to produce new medicine and other important technologies. Much of the early stage novel, risky, ground-breaking research only happens because it is funded by philanthropy. Large funding sources in government and business are risk averse and much more interested in supporting incremental but certain progress — yet the only reason that progress even exists in the first place is that someone was willing to put money on the very early stage work that made it possible. Similarly once a new technology is shown to be possible it is a good thing to aid deployment and continued improvements in implementation that help ease the hurdles to bring the results to less fortunate regions of the world. The sure things never do much; it is the radical new approaches that enable real progress.

Everyone gets to choose where they put their money and how they think about the world. Why are all those rich people funding heart treatments and better drugs for an age-related condition? Isn't that just selfish? Yet the distinct character of our era is that access to technology is comparatively flat: the progression of availability from expensive and inaccessible to accessible to the vast majority of people occurs very rapidly. Look at the spread of mobile phones and internet access over the past two decades as an example of what happens in a market where governments interfere far less than is done in medicine. Even in medicine, many of the medical technologies funded by rich people in past decades, and initially only available to the wealthy and connected in their earliest forms, are now available in places that include rural India and reaches of Africa — such as those drugs for heart conditions.

Improving medicine is not about making things happen for the wealthy. It is about whether we all win together or we all lose together. Mocking or shunning improvements aimed at preventing the suffering and frailty of aging because some wealthy person might get the treatments first is lunacy — that is what being wealthy gets you, and if it serves as an incentive to get them involved, then all the better.

Here is another way of looking at this: what causes the greatest harm to poor people? It isn't malaria. It is aging, and by a long, long way. Malaria killed in the vicinity of 650,000 people in 2012. In that same year somewhere north of 40 million people aged to death. More than three quarters of the world's population are exceptionally poor compared to the people we call poor in the US: so perhaps 30 million or more of those deaths fall into that demographic, fifty times as many as caused by malaria. I think it fair to say that degenerative aging places a far worse burden upon those individuals than on you and me. It is terrible for all of us, and kills all of us if we're lucky enough to evade the rest of life's slings and arrows, but there's a big difference between being old and frail in agrarian poverty versus a first world city. If you are a rational, compassionate, utilitarian individual — and few are, sadly — then it should be clear that the best thing that can be done with limited resources is to work as rapidly as possible to produce effective treatments for aging that prevent and reverse age-related disease. Just getting a first generation of these treatments into the development pipeline at all, and not even taking any further steps beyond that to help speed things along, guarantees that the poor of the near future will have far better lives as a result. We would hope to do more than just that, of course.

The greatest positive change we can create in the world is to eliminate the pain and suffering of aging through medical science. The outcome in terms of future lives saved and lives improved is so large in comparison to the treatment of any specific disease, even endemic diseases such as malaria, that it compels attention. That said, overall medical research funding is tiny in comparison to the wealth that flows through the entertainment industry, that goes towards killing people in ever more inventive ways, that is used to make candy, that changes the color of a US president's vest from red to blue, and so on. We like our wars and our circuses, and the scraps left over after that is done are all that goes towards making the world better by building new medical technologies. There is more than enough funding out there to cure every disease, to grow the life science research community by a factor of a hundred, and achieve countless other important goals besides. People just choose to spend it on other things, all ultimately pointless, forgotten, and irrelevant in the long term. The only thing that really matters is progress in technology, and especially in medicine, but persuading the world of that fact is still a hard sell.

GERONTOCRACY AND OTHER OBJECTIONS

Fight Aging!, 22 March 2013

Gerontocracy: government by a council of elders, or simply, government by the old.

There are many knee-jerk reactions to the prospect of greatly increased healthy human life spans. Some people believe that longer lives will result in stagnation, which is actually one of the more ridiculous and improbable outcomes once you start to pick it apart in any detail. Human society is restless and changeable on timescales far shorter than current life spans, and the reasons why are rooted in day to day human nature. Our ambitions operate on a horizon of a few years, and that wouldn't change all that much were we to live for centuries. We are driven to influence the world today, now, regardless of the years that lie ahead of us. So the fashions of this year are gone by the next. The idols of popular culture rise and fall with rapidity. The political and business leaders of this decade are gone in the next, displaced by peers. Even corruption and revolution on a grand scale are usually only a matter of a few decades, not lifetimes.

A subset of these beliefs on human longevity and stagnation involve the nebulous fear of a future gerontocracy, the rise of a self-perpetuating ruling elite of ageless individuals. Funnily, this is often voiced by people who are, unlike myself, perfectly comfortable with today's Western governments. I say funnily because I have to ask: are not our present societies already gerontocracies? Isn't any civilized society a gerontocracy? Who has had the most time to gather connections, a network, and make good use of them? The old. Who has had the most time to gather resources and invest them? The old. Who has had the most time to become truly talented and sought after? The old. Who has had the most time to work their way through a social hierarchy to challenge its existing leaders? The old. Where then will the elite and the leaders tend to arise? From the old.

Take a look at who just runs and influences companies, governments, knitting circles, successful non-profit initiatives, extended families. Young leaders exist, but they are a minority among the ranks of the old. This is the natural state of affairs for any society that possesses enough technology to make thought and craft more important than strength and vigor.

Even as power is centralized, there is still a year by year turnover of figures — even in the most defensible and corruptly secure positions of power and influence. They are largely kicked out by some combination of their peers and the mob in the sort of political anarchy that exists at the top, above the laws made for the little people. It is the rare individual who can stick it out long enough to be removed by the infirmities of age, even now, in this age of human lives that are all too brief in comparison to what is to come.

But back to the point. We live in a gerontocracy, and so did most of our ancestors. Yet change still happens just as rapidly as in past centuries when fewer people lived into later life in the sort of good shape they can manage today. Fear of some sort of comic-book gerontocracy emerging in the future seems, frankly, somewhat silly.

One io9 piece treated the fear with rather more respect than I'm inclined to give it, working through the case that an ageless generation of seniors would refuse to relinquish their grip on corporate boardrooms, capital, and political office, and pressing the futurist James Hughes on what happens if an undying elite simply won't let go. His answer undercut the whole premise: the trouble was never the chronological age of the people hoarding power, but the existence of small, unaccountable elites at all — and he doubted that longer life would leave anyone's mind any less flexible than it is today.

There are always outliers, but they are rare in comparison to the vast majority of leaders and the famous who are just part of the churn, coming and going, displaced and quickly forgotten once their few years are done. The top of a pyramid is a challenging place to stand.

Religion and the Long Life

Fight Aging!, 29 August 2014

There exists a very long history of people starting explicitly religious movements to achieve secular goals in the mundane world. A very broad range of such ventures can be found even in recent history, in this apostate age of comparatively weak and apolitical mainstream religious institutions. For every group who set out to achieve their ends through political advocacy, such as by starting a single issue political party, you'll find another who choose to start a church. Those who do this outside the established religious mainstream are looked at askance, and often deservedly so, but really you're just as likely to be taken for a ride on either side of the fence.

I want to glance here at a very narrow thread of religious ventures, fictional and otherwise, that winds its way from the science fiction of the golden age and later through satire and art of the 70s and 80s and then into the online transhumanist community of the 90s and its present descendant organizations. At all these points in time, someone thought it a plausible idea that a religious movement should exist for the purpose of preserving and propagating secular technological goals, in particular the extension of healthy life spans through progress in medical science.

As an aside, I should mention that when I'm out there talking to people about longevity science and the need for funding, you can be certain that I'm not bringing any sort of religion to the table. In fact I'd be happier if religious movements started for the purpose of promoting the secular goal of life extension didn't exist, and for many of the same reasons that I'd rather the more vocal "anti-aging" supplement sellers and snake-oil salesmen went away. To the extent that these people succeed in gaining attention for themselves, they make life harder for the rest of us. Nonetheless, these initiatives exist, they are a part of the extended community of supporters, there is a fascinating history bubbling behind it all, and therefore this is worthy of at least a short note on all of the above.

If you are widely read in science fiction, you will have no doubt encountered numerous tales in which the principals are engaged in what might be called religious engineering: attempting to change the world by means of exploiting the religious impulse as a lever, with varying degrees of cynicism and spiritualism involved. Ideas don't stay stuck in their books: they take wing when they are found attractive.

Starting a religious movement for reasons that are openly secular, as opposed to keeping your secular goals hidden behind a veneer of faith, has come and gone as a fashion over recent decades. You might look at Discordianism, the Church of the Subgenius, and even Pastafarianism as examples of openly secular religions. The degree to which people do take these things religiously is a matter for debate, but it is hard to argue against their success as cultural movements in their time. These are the children and the siblings of science fictional religious engineering, ideas for advocacy escaped and made real.

So given this promiscuously inventive history, we shouldn't be entirely surprised to discover small religious movements and proposals for the same associated with the support of longevity science. Any community of a meaningful size will produce such initiatives, and the modern community supportive of radical life extension achieved through advances in medicine is now several generations old, spanning a time of great invention in religious engineering, both inside and outside the mainstream. One of the first examples of this sort of thing that I recall encountering via the transhumanist community was the Church of Virus, a clear descendant of numerous science fiction concepts. That, perhaps appropriately, is little more than an idea: a proposed memetic framework meant to compete with traditional religions by offering adherents a path to a meaningful life and to immortality without recourse to mystical belief.

There are also populated groups, however, such as the Society for Venturism, which has been a co-traveler with the cryonics community for some time now: a minimalist secular religion built specifically around that community's practical needs — giving cryonicists religious grounds to refuse an autopsy that would ruin their chances of repair, and raising funds for patients who need last-minute cryopreservation funding.

Similarly, and of a more recent origin, there is apparently a Church of Perpetual Life, which has the sound of Unitarianism for supporters of radical life extension and cryonics: a non-denominational, science-based congregation open to people of all faiths and none, built around the goal of extending healthy human life and providing fellowship for longevity enthusiasts through regular services.

As it becomes increasingly known among the public that new approaches to treating aging and extending healthy life spans are on the way, I can only imagine that we'll see more of this. The large mainstream is always slow-moving and conservative, and as a consequence its leaders will find themselves challenged on the topic of extended healthy life spans, just as they have been challenged on medical advances in the past. They will fight a little, and then smaller and more nimble challengers espousing more sympathetic theologies will gain adherents, and ultimately the mainstream will accommodate them or be replaced: the same brand, but different bishops running the show. I'm sure it will be interesting to watch, for all that it has little relevance to presently important goals and challenges in advocacy for longevity science.

Mortality and the Shape of Society

Fight Aging!, 22 October 2012

Mortality determines society. By this I mean that the function describing mortality rate by age in humans acts as a filter to determine the sort of societies that can exist and support themselves. A high mortality rate reduces the range of societies that can exist. The mechanisms at play in this relationship are economic in the broadest sense: a mix of population growth, division of labor, effectiveness in recording and transmitting knowledge, time preference, establishment of sufficient capital for more than just subsistence living. Consider the difference between the late paleolithic era and classical Greek civilization, for example, and think on why they are different.

In terms of mortality, the immediately noticeable difference between the classical era and prehistory is an upsurge in the number of people who reached old age — a reduction in the omnipresent violence between small groups, and a marginal improvement in medicine. As a general rule you don't find the bones of old people in early human archaeology: so few of them made it past 40 as to be close to absent entirely as a class. Mortality rates in classical Greek and early Roman times were horrific by modern standards, but enormously improved over those of prehistory.

Thus as a consequence there were both old people and a certain expectation of reaching old age — which means a rise in planning ahead and an increase in available capital. A longer vision breaks through destructive cycles of economic behavior that take place on shorter timescales. As one example, people cultivate their property more effectively if they think that it must last for longer under their stewardship, more readily resisting short-term gains that come at the cost of long-term gains. This incentive has operated throughout human history, whenever average life spans rose. As farmers in 18th century England grew aware that old people were living longer, historians have argued, they grew more concerned with their own future subsistence — a plausible spur toward saving more, consuming less, and planning on a longer horizon.

Various destructive cycles of poverty to wealth to poverty exist for all time scales of human behavior and all population densities. A greater expected span of life doesn't make them go away entirely, but it does mean that more people will be around to suffer the consequences of ill-thought action — and as a result, endeavor to understand and avoid the acts that lead to suffering they see in others. Don't eat your own seed corn; be the ant rather than the grasshopper; don't cut down the whole forest; engage in trade rather than slavery; shun destruction in favor of construction; allow the merchant class to exist; and so forth.

One important thing to note is that all historical progress in expanding the range of possible human societies has come about without greatly altering the maximum human life span. It was possible to live to more than a century without modern medicine and in the presence of pervasive disease and violence — that outcome was just enormously unlikely for any given individual. Improved mortality rates led to a growth in the number of old people and their influence.

I submit that this largely unchanged outer limit to life span is why we can look back on thousands of years of politics to see the same patterns repeated over and again. Regions rise and fall. Empires form and decline. Countries become wealthy and then poor. Democracies slide into authoritarianism. Currencies are steadily debased and destroyed. Known and enumerated forms of governance arise over and again, and fail in the same ways each time. The patterns repeat because throughout history expectations regarding the outer limits of personal stewardship and responsibility remained set at something less than a century. It is that span of time that drives choices made between cultivating future growth versus squandering it for present advantage — and even drives whatever incentive exists to understand that you are taking one of those paths rather than the other.

After all, economics as a professional body of knowledge and theory only really evolved into the forms we'd recognize today in the wake of 17th to 18th century gains in longevity, after the rise and fall of physiocracy as a flawed means to explain the wealth or poverty of nations and how that related to human behavior.

Just as paleolithic hunter-gatherer groups were stuck within their own limited range of sustainable societies, so are we stuck within our own, larger range of possible ones. Our present span of life limits us to what we see. Where wealth, freedom, and the rule of law necessary to create wealth arise, they corrode over time as successive generations forget how and why their good fortune came about: they eat their own seed corn. We can see this happening in the US, somewhere past the mid-point of the process, and all somewhat analogous to the decline of the Roman Republic, at least insofar as root causes are considered: successful republics have a way of falling into empire and authoritarian rule, accompanied by massive military and welfare spending. Bread, circuses, and the legions.

It is the consequence of our dominant short-termism, where our time preference fails the test, where we squander our own potential.

The human condition has been bounded in this way for a very long time indeed, since the first cities arose more than 6,000 years ago, but it will not continue for much longer. In this age of biotechnology the transition from age-bounded life spans of a century to accident-bounded life spans of thousands of years will happen in the course of mere decades. Entire populations will stop dying as therapies capable of repairing the cellular and molecular damage of aging become widely available. Degenerative aging will become a controlled medical condition, warded and defeated like smallpox and others in the past century.

The wheel of time will be broken. Those who stand at the beginning and the height of empire will be the very same individuals who must also stand at its decline: their horizons will extend, and so the shape of empire will be quite different, if it comes about at all. How will this play out, this grand restructuring of all the incentives that lurk at the base of human societies? What new forms of society will emerge as possibilities given widespread radical life extension? These are open questions: the changes to human life span that lie ahead are dramatically different in character to those that have occurred in the past. The gulf in mortality and life span that likely lies between us and the humanity of the 22nd century is far greater than that separating paleolithic hunter-gatherers from the classical Greeks. Further, it will occur in a comparative eyeblink — a single generation will stand with one foot in each world.

Yet the wheel of time will be broken, and this seems to me to be a good thing.

Better Myths

Fight Aging!, 17 February 2007

"Every story is the story of the Fall" — except the one that matters, the one we're all writing together with quills of science, will and toil in the real world. That story is a grand arc of irresistible rise, of the defeat of obstacles and surpassing of limitations to our true potential. But you wouldn't know it from the myths that we find most comforting, as illustrated by their widespread nature.

I forget where I first heard the quote above: "story" in that context means "myth." Myths are not a remote line item from our history; they are the resonances in the conversation of ten thousand winding threads that takes place across a culture; common, retold and reinvented stories we tell ourselves to describe who we are, where we've come from, where we are going and what it is we want.

So it is that advocating and working towards healthy life extension is greatly affected by the stories we tell to one another about aging and its place in the world. I have written about this before: given a technology that allows us to do something new, we will turn that capacity to build our world a little closer to the world of myth that makes us comfortable. In the process, some feedback or change to the human condition slowly — very slowly — introduces changes into our mythic structure.

Myth drives the application of technological capabilities, which then in turn change the myth — but in the short term, myth is in the driving seat, all other things being equal. For our near-term future, the myths of longevity and immortality — and the myths used to make us feel better about lacking both — are important considerations when it comes to raising support for research to greatly extend the healthy human life span.

The common thread running through our earliest mythological tales of immortality is that death, while not necessarily desirable, is unavoidable and a fundamental part of being human. In spite of his best efforts, Gilgamesh could not achieve immortality. Given a choice, Tithonus likely would have chosen death over eternal decrepitude. While there is no suggestion in these stories that death gives value to life, they both certainly leave the impression that we should grow accustomed to the idea of dying, because there can be no desirable alternatives to it.

The story of the Fall is an old and simple one; the world is one of shortages, pain, suffering and death, yet we humans can conceive of a world absent these troubles. Nostalgia is a part of the human condition also — we see earlier times in our own lives as better than they were, and it's a short leap from there to draw a line of decay from an imagined golden age to the imperfect present. The Fall is an alignment of the mythic world — a better, imaginary world — with the arrow of time; for a variety of reasons, we have come to put that mythic world in the past rather than the future.

If you look for the Fall in subtexts far and wide, you'll see it everywhere. Malthusians, luddites and many environmentalists immediately spring to mind as those who most obviously buy into and support the myths of the Fall in the most modern way. To their view, the world is going to hell in a handbasket, and the myths of that Fall — for whatever reason — resonate most in their currents of the cultural river. But for every outright Malthusian you'll find a hundred young people who will claim they'd rather age and die than live to see the hand-me-down, grey future they imagine for themselves. They're not up for helping meaningful progress: why should they? That isn't the place for themselves they have defined, nor even meaningful in their view of the world. Such is the power of myth.

Sadly, one result of the comparative wealth and prosperity brought by modern technology is that ever more people are insulated from the consequences of being absolutely wrong about the way in which the world works for long enough for great damage to be done. Myths that, if followed, will lead to poverty and suffering can thrive for long enough to do just that.

We need better myths if we are to gather support for a major effort to extend healthy life span, analogous to the quest to cure cancer — these myths do in fact exist, but, unfortunately, are not widespread to the degree required. So long as the vast mass of people are entirely comfortable to describe a world in which aging is inevitable — and the role of the elderly is to suffer, become frail and die on schedule — then progress towards healthy life extension will remain frustratingly slow.

If people choose to reject healthy life extension, then so be it: but you have to first hear the myth in order to turn your back on it. Insofar as advocacy goes, we're still at the point today of telling our stories more loudly, and broadening the audience. Strive to live healthily and live long — we think you'll find it's a better class of myth for the real world than those in which the hero dies young.

PART IV

Why This Time Is Different

Most great ideas are brought into existence about 50 years after people stop considering it ridiculous.

— Fight Aging!, March 2004

Longevity science has a credibility problem, and he knows exactly why: the field has a long history of hucksters, and he can smell one at a distance. These chapters make the case that this time is different, explain how real progress actually arrives, and show him policing his own side as sharply as he attacks the other.

The Fight Over "Anti-Aging"

Fight Aging!, 11 March 2004

There is a war being fought over the meaning of "anti-aging" and (to a lesser extent) "life extension." It's fought with words and funding between and within a number of different factions within the business and scientific community.

A recent skirmish in this war — the Silver Fleece award made by Jay Olshansky to A4M at the International Conference on Longevity — was a fairly typical exchange of views, but involved some of the players in the wider war who are of more interest to advocates for serious research into understanding and curing aging.

On the one side, we have Jay Olshansky, regarded as one of the more conservative gerontologists. He has said that radical extension of the healthy human life span is impossible in the near term, and maybe also in the long term. This is an extreme position that Jay Olshansky must defend within his own field. Moderate gerontologists like Stephen Austad (with whom Olshansky has a well-known public bet on healthy life extension) and forward-looking biogerontologists like our own Aubrey de Grey hold that large gains in healthy longevity are possible within our lifetimes.

On the other side, we have the American Academy of Anti-Aging Medicine (A4M), often represented in the media by Ronald Klatz. I will admit to being confused by the actions of A4M; they seem to straddle the new school and old school worlds in the community. For example, A4M was an early donor to the Methuselah Mouse Prize and Ronald Klatz speaks out in support of therapeutic cloning, stem cell research and regenerative medicine as tools that will extend the healthy human life span. On the other hand, A4M runs anti-aging conferences that are widely reviled in the scientific community due to the presence of vendors from the fraudulent and adventurous marketing school of "anti-aging" pills, potions and crystals. At the same time, Ronald Klatz has expressed the desire to see the "marketeers" vanish from the industry, and recognizes the great harm they are doing to scientific progress.

Olshansky views A4M as doing great damage to legitimate research and science. The founders of A4M view Olshansky as an entrenched member of an establishment unwilling to grant legitimacy to any of the "anti-aging" industry, no matter what evidence is offered or views espoused.

This war is being fought over money, but money takes second place to the perception of legitimacy. It is this perception of legitimacy that determines funding for scientific research and revenues for businesses. Scientists feel, quite rightly, that the noise and nonsense coming from the "anti-aging" marketplace is damaging the prospects for serious, scientific anti-aging research. If everyone knows that anti-aging means high-priced cream from Revlon marketed to the gullible and brand-aware, no scientist is going to get funding for a proposal that uses the word "anti-aging." Worse than that, people start to assume that real efforts to reverse aging must be impossible.

Businesses in the "anti-aging" marketplace make money from the aura of legitimacy whether or not their products perform as advertised. Those businesspeople who do feel they have working products carry out their own fight against opportunists, frauds and "marketeers" who they feel are damaging their own market. Ironically, this is much the same argument used against these businesses by scientists like Jay Olshansky. The vast amount of money spent on products that claim to turn back the clock demonstrates that people want real anti-aging medicines. The trouble is that these real anti-aging therapies simply don't exist. Or do they? It all depends on how you define "anti-aging."

I've spoken before about "optimizing natural longevity" in the context of trying to draw a distinct line between what you can do now to lead a longer, healthier life, and what will be possible in the future. We can presume that there exists, for each person, some maximum life span that you can reach using the technologies and understanding of today. You can adopt calorie restriction, exercise, keep a good relationship with a physician, and spend an appropriate amount on supplements and healthcare. Each of these items will help you to live longer and in better health than you would otherwise have done. Does this make them "anti-aging," preventative medicine, good maintenance, or merely not damaging yourself quite so much?

If an improved supplement comes onto the market that adds a few years of life through some biochemical mechanism, is that "anti-aging"? All of these things are clearly going to extend healthy life span by some amount. We could spend a lot of time arguing one way or another (and proposing further, more ambiguous examples). When Ronald Klatz says "anti-aging," however, I'm fairly sure he doesn't mean the same thing as Jay Olshansky's definition.

One problem is that we don't have any way of measuring effectiveness for a proposed anti-aging treatment, short of waiting for the subject to keel over. This is clearly not the desired experimental approach for those of us who want to see radical life extension in our lifetimes. We need biomarkers for aging: ways of measuring the progression of the aging process in our bodies. Even if we do find aging biomarkers, however, it isn't clear that they will allow accuracy to the point of being able to say "this treatment is giving you an extra two years of healthy life." Car enthusiasts can tell when they're getting that last 10% out of the engine, but you can't determine that sort of thing when examining health and life span.

I'll leave you with this thought to mull over: if we possessed medical technologies that could extend the healthy human life span to 150 years (or more), I think it's a fair bet that we wouldn't be arguing about the semantics of anti-aging and life extension. In large part, this money-fueled argument is entirely due to the absence of working anti-aging medicine that can greatly extend our healthy life span.

Why Now Is Not the 1970s

Fight Aging!, 21 January 2015

Every culture throughout recorded history had its seekers after agelessness, all of whom were deluding themselves. As science replaced alchemy the seekers remained just as prevalent, but adopted the superficial trappings of science in their futile quest. A few even adopted the scientific method, or emerged from the scientific community of the time, and were thus much more rapidly and reliably disappointed by the results of their experiments. The power of the scientific method lies as much in its ability to close off potential paths ahead as to open up new ones: it clears out wishful thinking and delusion for those willing to adopt its rigors.

Technology and other applications of scientific knowledge have steadily lengthened healthy life spans since the late 1700s, once the positive feedback loop of growth in wealth and knowledge really kicked in — slowly and incrementally, adding roughly a year of life expectancy at 60 with each passing decade.

In the 1970s futurists were very enthused about the prospects for medicine, and especially in the prospects for their own personal longevity. They are all aged to death or near as gone now. They were absolutely wrong about how much could be achieved with then new and exciting applications of biotechnology. Yet so very much has been achieved. In comparison to the tools of today, 1970s biotechnology is clunky and expensive: halls of manually tended machinery have now shrunk to a single chip, and a graduate student today can accomplish tasks in a few weekends that would have strained the largest laboratory in the country for years back then.

So we're all pretty excited about what can be done today in medicine, and the prospects for our own personal longevity. When it comes to our understanding of biochemistry and ability to manipulate our cells, we are as far beyond the 1970s as the 1970s were beyond the gentlemen-scientists working at the end of the 19th century. Why, however, is it different this time? Why are the seekers after agelessness now rational scientists rather than another crop of self-deluded fools? This is a question that crops up. I can recall numerous conversations over the years in which I was informed that someone knew an older fellow who was, back in the day, quite confident in the forthcoming existence of longevity-enhancing therapies, and yet where are those treatments decades later? Nowhere in evidence, but here I stand telling you that now is the time, that the Strategies for Engineered Negligible Senescence (SENS) are a viable, plausible road to rejuvenation treatments that could indefinitely extend human life, and that given sufficient funding we could make enough progress in the next 20 years to hit actuarial escape velocity, the point at which medicine adds more healthy life faster than aging takes it away.

So why is it different this time? For one there is SENS, a detailed plan of development leading to rejuvenation treatments that could be prototyped in mice given a billion dollars and ten years, give or take. No such plan could have been formed a century ago; SENS could not have been proposed as a serious project back then, even had someone had the realization. There was simply no way to even guess at how much time and money it would have required to build the tools to build the tools to develop the validation of the theories so as to build the tools to build the tools to develop the therapies, and so forth: it would have been a project on the scale of going to the moon, and with far less certainty of success.

None of the proposed paths to add decades or more of healthy life put forward in past generations, now obviously naive and wrong, were remotely as rigorous, or as broadly supported within the scientific community. Only now do we have that, built on the vast body of knowledge of biology accumulated over the last century, and on the new tools of biotechnology of the past few decades. Only now are large numbers of scientists putting their careers and their reputations into the extension of healthy life.

Why is it different this time? Funding for various scientific establishment efforts to extend life is growing rapidly. Most of these are in fact not going to move the needle all that much, but that isn't the point. The point is that the consensus in a significant fraction of the scientific community and its surrounding institutions of funding and review is that the time has come. Scientists are the arbiters of truth in our culture, though this is sometimes hard to see, and the rest of the world will follow their lead when deciding whether to take something seriously. That will create a feedback loop of funding and progress in which, yes, a lot of less useful work will thrive, but so will significant approaches such as SENS.

None of this was the case for past generations of what turned out to be deluded optimism. It is the case now. The times have changed, and it is different this time around.

Nobody Says It's Impossible Anymore

Fight Aging!, 13 August 2011

A thought for the day: nobody out there is seriously arguing for the impossibility of radical life extension, and I don't think anyone has been for quite some time. It is a given in the present diffuse discussion on the future of medicine and human longevity that at some point advances in biotechnology and nanotechnology will lead to greatly extended lives: centuries and longer lived in good health and vigor. Aging will be brought under control by medicine, like any number of other once intractable medical conditions.

It wasn't always this way. People in past centuries might have hoped for the plausibility of radical life extension, but couldn't have said in certainty that it was possible. We, on the other hand, know far more about physics, chemistry, and biology: we know that there is no wall created by the way the universe works standing in our way. The only reason we presently age and suffer is because we haven't yet advanced far enough down the path of biotechnology that is clearly visible and well understood. Aging is, at root, a matter of atoms and molecules in the wrong place and the wrong configuration. Moving atoms and molecules around to order, en masse, and with precision, is a task that we know is possible.

Yesterday the tools were found molecules that happened to do something useful with other molecules. Today we make use of designed molecules for particular operations, knowing much more about the molecular machinery of our cells. Tomorrow the biotechnologists will build and repair complex molecular machinery that performs far more effectively than our evolved biology.

Thus discussions on the engineering of human longevity focus on how, when, and (sadly) whether it should be done at all. I see great strategic importance in the right groups gaining ground in the "how" discussion — we're all going to age to death just like our ancestors if the scientific community remains focused on metabolic manipulation to slow damage accumulation rather than the repair of damage exemplified by SENS, for example. Similarly if there are not good inroads made in growing the community of researchers interested in SENS and related lines of research.

Discussions on "when" can probably be skipped as lacking rigor: no one knows. All the meaningful timelines depend greatly on seeds sown now that will only bear fruit in the 2030s — the course of twenty years remains a matter of long-term planning and great uncertainty in specific outcomes while we're stuck living lives that top out at a century (and that with great luck). It is interesting to ponder and plot the windings of future events, but that time is probably better spent on influencing the "how" discussion or materially contributing to progress.

As to the discussion on whether engineering longevity is desirable, or should be blocked by people in power, I think it never hurts to take a little time to oppose such lines of thought. That opposition is widespread, and hard to shift.

But all that said, no one out there is seriously arguing that radical life extension is impossible.

It Is 1885 and We Are Arguing About Flight

Fight Aging!, 15 August 2012

The cautious majority believes that human life span will continue to increase, but only incrementally, much as it has done for the past few decades — both life expectancy at birth and life expectancy after 60, due to the continued introduction of new medical technologies (which proceeds far more slowly than it could, thanks to the heavy hand of the state).

To my mind, arguing for incrementalism in any trend relating to medicine at the present time is choosing to go against the tide. The biotechnologies that underpin advances in medicine are going through a period of massive, revolutionary change. While it is true that organizations such as the FDA do pretty much everything short of shooting scientists to slow down and increase the cost of turning research into therapies, the rapid pace of progress in the life sciences will win through.

Allow me to put forward a historical analogy: standing in 2012 and arguing a case for gentle future changes in life expectancy over the next few decades, based on the past few decades, is something like standing in 1885 or so and arguing that speed and convenience of passenger travel will steadily and gently increase in the decades ahead. The gentleman prognosticator of the mid-1880s could look back at steady progress in the operating speed of railways and similar improvement in steamships throughout the 19th century. He would be aware of the prototyping of various forms of engine that promised to allow carriages to reliably proceed at the pace of trains, and the first frail airships that could manage a fair pace in flight — though by no means the equal of speed by rail.

Like our present era, however, the end of the 19th century was a time of very rapid progress and invention in comparison to the past. In such ages trends are broken and exceeded. Thus within twenty years of the first crudely powered and fragile airships, heavier than air flight launched in earnest: a revolutionary change in travel brought on by the blossoming of a completely new branch of applied technology. By the late 1920s, the aircraft of the first airlines consistently flew four to five times as fast as the operating speed of trains in 1880, and new lines of travel could be set up for a fraction of the cost of a railway. Little in the way of incrementalism there: instead a great and sweeping improvement accomplished across a few decades and through the introduction of a completely new approach to the problem.

This is one of many historical examples of discontinuities in gentle trends brought about by fundamentally new technologies. Returning to the medicine of the present day, there are any number of lines of work we could point to as analogous to the embryonic component technologies of an aircraft in 1885. They are still in the lab, or only being trialed, or still under development — but they exist in great numbers. There are the SENS technologies, various forms of advanced immunotherapy, and other approaches besides. Just because we can't see the exact shape of the emerging technologies that will be constructed atop these foundations doesn't make them any less likely to be created: great changes are coming down the line in medicine. The future is not one of steady and incremental progress.

Biotech's Cost Curve

Fight Aging!, 21 January 2014

Drawing on historical analogies is useful for figuring out where we are and where we're going: human nature doesn't change, and the basic organization of research, development, and commercialization still resembles that of a few decades back, closely enough that recent history is safe to mine for examples.

Biotechnology is the application of the life sciences, and the foundation of medicine. At present progress in biotechnology is rapid and revolutionary. The groundwork for a series of disruptive, factor-of-ten improvements in a variety of medical technologies has already been accomplished: think of the attention given to gene sequencing over past years, for example, the world watching as costs plummeted even while capabilities increased year over year. That is just one of many, many applications in biotechnology that are improving in similar ways.

This present stage bears real resemblance to the dawn of powered flight — decades of component technologies developed in isolation or assembled into noble failures, then a sudden leap that, within a few further decades, revolutionized the whole nature of travel.

A different sort of leap occurred over the 1960s and 1970s in the development of modern computing: the move from expensive, large computing devices to cheap, small computing devices. A mere change in price is far from prosaic and boring: it drives sweeping changes in adoption and expansion in the forms of application for any given technology. The lower the barrier to entry — price in this case — the more experimentation and thinking takes place, leading to greater application of a given technology for the benefit of all. In the past I've pointed out the parallels between the early personal computing societies of the 1970s and the present DIYbio community: enthusiasts, hobbyists, and professionals merging their efforts as costs fall to the point at which anyone can join in and build.

The Computer History Museum's timeline entries for 1960, 1965, and 1970 make it not unreasonable to suggest that we're somewhere in the middle there when it comes to biotechnology today. The present trend I have in mind is the move towards small, portable, low-cost laboratory equipment that can accomplish most of what was possible in a large lab ten years ago. A lab in a box by that metric is years away still, but numerous groups are making significant inroads towards that goal.

But why care about this? In truth it isn't all that important to me whether or not 2010 in biotechnology is 1960 in computing, but it is important to me to have some model for the next few decades of work in medical science, and in particular the ability to make progress towards treatment and reversal of aging. How much progress is likely in the foundations, the capabilities in biotechnology? Should we expect and plan for disruption of the research process to the point at which the barrier between trained professional and effective self-educated contributor blurs to nothing, as has happened for software development?

We can ask these and other questions so as to have some idea as to whether matters are proceeding as fast as they might, and what new directions in effective advocacy and funding will arise in the near future. Clearly at some point if the institutions are not funding the research and development work we feel is important, it will be possible to do it ourselves. Given smart, motivated collaborators and crowdfunding, complex problems can be solved in software development today. Creating applications of new knowledge in molecular biology and medical science is no more complex than constructing big software projects, though regulation obstructs it far more, and the tools remain expensive and anemic by comparison to what software developers enjoy. The latter will change, rapidly, and the former will simply mean that development will occur most readily in regions outside the US.

But these are things to think on as we watch and support our favored research into aging and longevity. In past decades we could only have watched — the amounts of money involved were too large for any ordinary individual to help with. But the cost of life science research is plummeting, and here and now crowdfunding of research projects that advance the state of the art is a very real thing. Medical research for your favored causes isn't just a spectator sport anymore: the times are changing.

Against Kurzweil

Fight Aging!, 21 September 2005

I have been working my way through Ray Kurzweil's "The Singularity is Near" (TSiN) over the past few days, having been the fortunate recipient of a review copy. The book might alternatively be titled "The Modern Futurist Consensus: a Review" or "Damien Broderick's The Spike: the Extended Remix." Those of us who have haunted transhumanist enclaves in the past few years (or more) are unlikely to find new ideas here, but the book serves a most useful purpose in bringing the best and brightest of transhumanist, futurist themes and thinking all together under one roof, in a popularist manner, with a unifying, easily-marketed theme. It's been done before — by the aforementioned Damien Broderick, amongst others — but not quite as comprehensively. This sort of book is something of a necessary precursor to wider advocacy and education in today's culture; a pleasant irony, given the subject matter.

My own two cents thrown into the ring say that the class of future portrayed in TSiN is something of a foregone conclusion. It's quite likely that we'll all be wildly, humorously wrong about the details of implementation, culture and usage, but — barring existential catastrophe or disaster — the technological capabilities discussed in TSiN will come to pass. The human brain will be reverse engineered, simulated and improved upon. The same goes for the human body; radical life extension is one desirable outcome of this engineering process. We will merge with our machines as nanotechnology and molecular manufacturing become mature technologies. Recursively self-improving general artificial intelligence will develop, and then life will really get interesting very quickly. And so forth — the question is not whether these things will happen, but rather when they will happen — and more importantly, are we going to be alive and in good health to see this wondrous future?

My criticisms of TSiN center on the timeline predictions for new technology, the pace of discovery, and above all the management of complexity. Arnold Kling made this point well in his own commentary on TSiN: progress toward general artificial intelligence has been slower than the futurists hoped, and I think that's one symptom of a broader underestimation of complexity management as such. The 1990s dreamers who imagined swarms of nanorobots assumed the hard part was building them; it turned out the hard part was controlling them. Replace "nanorobot" with "human cell" and that's where we stand today in biotechnology. Our bodies are immensely complex, and the reason researchers can make headway with gene therapies and stem cells is that they are, effectively, tweaking settings on machinery that already handles most of its own complexity management. The leap required to go further — replacing whole biochemical systems with artificial substitutes, or reverse-engineering the brain outright — is a much larger one.

Kurzweil's commentary on types of complexity in TSiN is a good read — and one of the better explanations for the layman I've seen — but it seems a little disconnected from the actual business of dealing with complexity in ways that matter. My take on it all is that science is largely the process of discovering keys to complexity; by this I mean finding algorithms, recipes or methodologies that enable us humans to understand and manage complexity that would otherwise be beyond us. To take an applied example, manipulating stem cells through comparatively simple procedures enables scientists to perform tasks — the regeneration of age-related tissue damage — that they cannot even monitor in detail, let alone control.

Humans building tools to interact with ever-greater complexity is nothing new, but our own, unaided capacity for managing complexity isn't increasing along with it — the keys we build have to be usable at our level. That process of building keys is what's accelerating, growing exponentially as Kurzweil argues, and we're only now leaving the stretch of the curve that looks linear.

There is one important area of complexity management in which we seem to be making little headway, however: the organization of humans in business and research. For all that we can now accomplish with faster computers and enormous leaps in telecommunications, we don't seem to have made significant inroads in getting large numbers of humans to cooperate efficiently. As Arnold Kling points out, that the excessive use of Who Moved My Cheese? is even in the running as something to try is not a good sign. I've been involved in a number of technological attempts to improve efficiency in large organizations, and the state of the art is not pretty — nor especially effective in the grand scheme of things.

I am prepared to go out on a limb here, as I have done before, and say that business and research cycles that involve standard-issue humans are incompressible beneath a certain duration — they cannot be made to happen much faster than is possible today. You can make such a cycle cheaper, but not much faster: strip away the technology and what's left is meetings, fundraising, decisions, and plain organizational friction, a human element that has proven remarkably resistant to improvement.

That doesn't mean these cycles can't be made cheaper, only that cheaper doesn't mean faster: more parallel attempts get made, more amateurs join the professionals, and any given problem is more likely to be solved within a single cycle of innovation rather than by cutting the cycle short. Biotechnology is not too many years away from this state of affairs, following the trail already blazed by software development.

Back in the deep end, expensive projects mean conservative funders, and conservative funders mean organizational matters proceed at a slow pace: blindingly fast rates of technological advance once the money is on the table, but the cycles of business, fundraising, and research still chained to the old human timetable. I regard this incompressibility as a real limit on exponential growth, one we are now hitting up against.

Kurzweil's Singularity is a Vingean slow burn across a decade, driven by recursively self-improving AI, enhanced human intelligence and the merger of the two. Interestingly, Kurzweil employs much the same arguments against a hard takeoff scenario — in which these processes of self-improvement in AI occur in a matter of hours or days — as I am employing against his proposed timescale: complexity must be managed and there are limits as to how fast this can happen. But artificial intelligence, or improved human intelligence, most likely through machine enhancement, is at the heart of the process.

Improve our capacity for handling complexity directly, and the old limits get pushed outward: we stop needing to search for keys to complexity and start managing it directly, and business cycles built around the limits of present-day human workers can finally begin to compress.

Until we start pushing those limits, though, we're stuck with the old organizational friction and the slow progress of general artificial intelligence — which is why I think Kurzweil is optimistic by at least a decade or two.

So how does this all fold into healthy life extension? Well, physical immortality is one obvious product of singularity-level nanotechnology, biotechnology and complexity management. There are no known barriers in physics to the construction of nanomedical systems capable of simultaneously managing, repairing — or replacing — every cell in our bodies. Even something as complex as the sum of all your cells can in principle be kept in the best possible shape for as long as you like — "all" it takes is knowledge, the future tools of nanoscale engineering and powerful enough computers. But when do we get there? This is the question, and it is one that shapes the actions of futurists and transhumanists. There are many who believe that the best sort of activism and advocacy for the future — even for healthy life extension — is in the area of artificial intelligence, because making self-improving intelligence arrive earlier will lead to all other currently pressing problems, such as age-related degeneration and death, being rendered trivial in the mid to long term. Obviously, I'm not in that camp: I'm sufficiently dubious about Kurzweil-like timescales — based on my views as set forth above — to think that we need to be tackling the problem of aging first.

Science Is Not Engineering

Fight Aging!, 7 January 2015

Technology is the application of scientific knowledge. The scientific culture and scientific process as it is practiced today embodies a strong bias against any sort of shortcut towards the production of technology, however. If it seems plausible at a lesser level of understanding of a system that you could achieve some beneficial application, then the peer pressure in the scientific community is always to hold off and work instead towards a full understanding. This situation is not uncommon in medicine: many discoveries are serendipitous, but to try to turn demonstrated positive results in the laboratory into positive results in the clinic will be opposed at every turn until the underlying mechanisms can be fully explained.

The bias against action and towards understanding as the primary goal is baked into every level of the research establishment and surrounding institutions. The scientific method has researchers moving eternally towards greater understanding. It has nothing to say about what you do with that understanding, however. Application to produce technology is where you step from the Platonic ideal and into the messy real work of engineering: the art of creating meaningful solutions in the absence of full understanding. This boundary between knowledge and technology produces all sorts of cultural friction, and I think it is fair to say that scientists who depart to be engineers are not treated as well as they might be by their former peers. There is little disapproval in the world quite like that of pure scientists directed towards one of their own who steps away to start a technology business.

At root there are very good reasons for this bias: the scientific method is required for progress, yet it is constantly under attack from opportunists and fallible human nature: wishful thinking, the desire to find progress where there is none, the desire for short-term gain over long-term gain, and so forth. If shortcuts are not treated as heresy, then there will be all too many research programs led astray, and development predicated on false results, and in the worst cases outright fraud. This happens even with the scientific culture of disapproval, but far less so than it otherwise might.

While the core of the scientific method is centuries old, to some degree this strong bias against early attempts at practical application is a phenomenon of modern times. In the institution of medical research everything that moves beyond early stage exploration must be explicable to the standards of the day, but matters were not always so rigorous. There are any number of grandfathered treatments available today that are not sufficiently well understood at the level of molecular biology to get into clinical trials if they were discovered today. Regulation compounds this: everything not explicitly permitted is forbidden, so funding and effort — even back down the chain to supposedly unconnected free-ranging fundamental research — collapse onto the molecular biology of late-stage disease and its proximate causes. For most researchers that's the only work with a future if you want to contribute to something that might actually end up approved for clinical use.

So these influential edifices of thought, and all the funding that is influenced by them, say that any approach in research that specifically aims to circumvent our lack of knowledge in some areas is simply Not The Done Thing. Yet there are many lines of research in which it is clearly plausible that great benefits could be derived by doing just this, and — as currently constructed — the walls that the scientific institution is forced to erect for its own defense exclude all of this good, solid work that might lead to better therapies. As one example you might look at leukocyte/granulocyte transfer therapies for cancer. These came to prominence on the basis of very promising results in animal studies, but have not seen significant funding or much work on human studies precisely because researchers cannot yet present a full accounting of how these treatments work. Without that, you won't see much movement.

In this context we of course come to the Strategies for Engineered Negligible Senescence (SENS). This is not just a proposal for the shortest path to the best results when treating aging, a research plan to create rejuvenation in the old and prevent all age-related frailty and disease, it is also a critique of the scientific community and its way of doing business. The present system is broken by virtue of the fact that its members have gone too far in building defenses against failure modes in the scientific method. They now systematically marginalize useful endeavors aimed at the production of meaningful results in absence of complete knowledge of the biological systems involved. Application of partial knowledge can be good engineering, and is viable, necessary, and needed in medicine, where every delay costs lives. The past shows that the engineering approach can be perfectly workable, as many drugs in use today were brought into use through exactly this sort of methodology, and their full mechanisms are in fact still not understood.

Still, the mainstream of the aging research community will continue to spend billions on efforts that have no greater expectation of practical utility than the sirtuin research of the past decade. They are working towards a full understanding of the overlap of metabolism and aging as a primary objective. Slowing metabolism is no substitute for repairing the damage it produces, and I object to its being presented as a serious near-term path to added healthy years. Researchers involved tend to think that adding five years of healthy life by 2030 is an ambitious goal. If that five years is all that happens, and when it does it certainly won't be five years for those people already old when treatments arrive, then what a waste of opportunity that would be.

No, we want to see work on rejuvenation, on ways to reverse aging by repairing its causes — work after the SENS model that has a clear plan to produce results in absence of a full understanding of metabolism throughout aging. That list of damage already exists; researchers don't need to understand every detail of how it interacts and progresses, only to repair it periodically. In this environment drawing more researchers to work on SENS repair biotechnologies is a bootstrapping process of attention and funding and results, just like the disruption of any industry by new technologies and new approaches. Things like clearance of senescent cells seem like one of the areas of research that will eventually get grudging attention by virtue of the fact that it will work far more effectively than anything the mainstream is producing with their drugs and their messing with the operation of metabolism in the late stages of disease. If you are trying to make a damaged engine work slightly better when damaged, rather than trying to address the damage itself, why would you expect good results?

The replacement of the present mainstream culture of aging research by SENS-like research focused on periodic repair of the causes of aging to achieve rejuvenation will proceed gradually and through the demonstration of effectiveness. There is a lot of replacing to be done, however. It's a long road yet, and conservative institutions will continue to support work that does nothing but add knowledge of the fine details of metabolism long after rejuvenation research is much more of a going concern, I'd imagine. Groups like Google's Calico venture do not fund SENS because they are run by exemplars of the current institutions of medical science, who have worked their entire careers in the world of full knowledge as a requirement and where the primary strategy is a struggle to alter metabolism in the late stages of age-related disease. Why should it be any surprise that what they are going to do is simply a continuation of existing programs of aging research? Calico will fund SENS at the same time that other institutions of the mainstream are doing so as a matter of course.

The Arcane World

Fight Aging!, 8 October 2015

"Arcane" is one of those words sorely abused by generations of people involved in the most fanciful of modern pastimes, which is to say the creation of alternative magical and religious beliefs and practices, both in earnest and for fun. As a consequence it has gathered a broad wake of connotations and cultural baggage. Cut all that away and it has an unbiased, simple, and straightforward meaning, however. To be arcane is to be obscure, to be hidden, to be known only to a few.

We humans have evolved a strong urge to pattern recognition. It is a key part of our intelligence as it applies to the business of surviving the rigors of natural selection. How else to explain magical thinking, the tendency for people to fit everything they observe into simple models of cause and effect regardless of accuracy, and then search for the truth later, if at all? The world about us is so very complex that there will always be things that any given group of people cannot understand or model accurately with the resources at their disposal, but put in that situation they will build the models anyway, because that is more comfortable than acknowledging ignorance. So we have religion, magical traditions, the ridiculous marketing that emerges from the snakeoil "anti-aging" marketplace, and generations of people attaching extra saddlebags of meaning to workhorse words such as "arcane," "esoteric," and the like.

Every present culture has very deep roots, stretching back at least centuries into eras in which it was universally accepted that an arcane world lies underneath the mundane, steering it, providing the rules that make sense of what seems senseless. In search of patterns, any patterns, people looked for guidance to whomever was bold enough to claim to see into the arcane world, and since we are a hierarchical species, like our fellow primates, that form of leadership was institutionalized into power structures very early on. Thus we have a history of shamans, priesthoods, temples, the Hero's Journey, and the endless, ever more baroque theology that commenced as soon as things advanced to the point of fighting with words and concepts rather than weapons. All of that lies underneath the thin veneer of modernism, a bone mountain, the legacy of the dead.

Now, here is an interesting thing about modern science and technology: its complexity and importance has in effect created a real arcane world that lies alongside the mundane, steering its future, determining who will live and who will die, what changes and what persists, how the rules of everyday life will differ tomorrow. The present state of technology is the greatest determinant of how we live our mundane lives, and technological progress is the greatest determinant of what tomorrow will bring.

Medicine and medical research, especially into aging, shape the rules that will determine the portents for the rest of our lives. How long will we live, will we suffer, what must we do to have the best odds of success? Two thousand years ago people went to priests and burned offerings. A thousand years ago they petitioned physicians who had more in common with priests than with today's practitioners. Today they go to clinics and understand about as much of the underpinnings of what they are told to do, for all that it is a lot more effective. The behaviors and organizations laid down to deal with the imaginary arcana of mysticism and religion continue for the real arcana of technology. Very few people go beyond talking to researchers to lift the veil and seek to understand why medicine is the way it is, why the answers to their questions are what they are. They accept the patterns that are explained in shorthand, and are comforted by them, right or wrong. That the patterns offered are better and more effective because of the changing tides of the arcane world of medical research is almost beside the point.

We who look further, with a foot in the arcane and a foot in the mundane, easily forget how distantly removed from all this we once were, or how much of an accident it is that we are where we are today. Ponder just how little thought you gave to medicine and where it came from when you were young, immersed in the mundane: back when you thought aging was set in stone, and the sum of the world was school, shopping, relationships, the changing of the seasons, a job, a hobby. The sum of an unremarkable, unique life. That is most people, unaware of what actually sways their futures.

All of this is why you see similar patterns of human organization at the high level emerging at the boundary of medical science and the world at large as at the boundary of organized religion and the world at large. The data is vastly different, and the importance vastly different. But the same underlying incentives and facets of human nature are at work, driving the small decisions that snowball into organizations and initiatives. For preference I'd like to see this change. The arcane world of medical research, and particularly that related to ending frailty and disease in aging, cannot continue to be as arcane as it is today if we are to see the growth we need in funding and support. The scale of applied resources and pace of progress that is justified by the grand panoply of suffering caused by disease and aging is hard to sustain when no-one thinks about medicine until they are sick.

If we could just bootstrap medicine to much the same position in the public eye as the automobile or the personal computer, where there is some breakdown of the veil between the arcane world of progress and development and the mundane world of use, even that would be a great gain. Unfortunately doing this is an uphill battle against our own evolutionary history and evolved preferences: threatened by complexity, and worse, by the time needed to make a dent in that complexity, most people retreat and direct their limited attention elsewhere. It is a hard sell to persuade anyone to outlay their precious time to understand something that will be important a decade or two from now. In effect those of us closer to the inside of the veil of the arcane for medical research are something like reverse Cassandras: knowing that wondrous, golden futures lie ahead, if only people will listen, understand, and help. More are taking notice with each passing year, but it is still far more slowly than they might. Changing the world is not easy.

How to Read a Science Story

Fight Aging!, 16 June 2009

The trouble with research and development in biotechnology and applied life science is that it's ridiculously complicated. Here I am, ten years into the forest as an observer of one small part of the field, and it's taken me this long to figure out how not to walk into the biggest trees or otherwise make a fool of myself. It's broad, ever-changing, and hard.

Now consider this: in any rapidly changing field of research there are facts and there are claims, where a "fact" is just a claim with very solid backing. It can be difficult for an observer, such as a journalist, to tell the difference between facts and claims. You interview some guy, or you're reading a few popular reviews, and they present a variety of information: some good, some bad, some pushing an agenda, some just wrong. Being published in a complex field is not a guarantee of correctness or of support for your thesis. If you're a journalist writing to a deadline for the normal pittance, for an audience who know even less about the field than you do on average, then why take the effort needed to pick out fact from claim? It just doesn't pay anything back to your bottom line.

This economic picture is why most popular publications on science are riddled with errors — checking fact against claim simply doesn't pay.

What prompted this line of thought was a recent piece at h+ Magazine on life span, antioxidants, and in particular the antioxidant superoxide dismutase (SOD) that is generated in various forms within the body. Frankly the piece is a mess, and a very good example of the false impression of a field you can end up with when you don't take the time to pick apart fact from claim.

The short version of my commentary here is that you, the reader, should take the implied conclusion of the piece — that manipulating SOD by means of pharmaceuticals is important to future engineered human longevity — as falling somewhere in that nebulous space between a lie, a red herring, a big unknown, and an unlikely outcome. More detail is in the following points.

The article paints SOD as important to longevity. There is by no means a weight of evidence linking increased SOD with increased longevity between species, however, or even within a species. If you even superficially survey the literature you'll find examples of varied forms of SOD overexpression, underexpression, and deletion moving life span in either direction under various circumstances.

The author further claims that declining antioxidant levels are an inevitable, "programmed" part of aging, and that drugs or supplements able to hold S.O.D. and other antioxidant levels at youthful levels are the only alternative to calorie restriction.

Now to start with, a brief survey of the literature tells us that natural antioxidant levels do not uniformly decrease with age in mammals. Some types of SOD increase in some tissues in response to raised levels of oxidants and oxidative damage, for example, even as other types fall or level off. But even taking what is written at face value, the conclusion is a huge leap from a known correlation to an assumed causation (less SOD is bad), and then again to an equally assumed single course of action (drugs and supplements). I'm all for speculation, and then testing on the basis of that speculation, but this is phrased as certainty.

The article makes no mention of the issue of location in the cell. A cell consists of thousands of distinct components, and demonstrated extensions of healthy life span in mice via the application of additional antioxidants show that the location of those antioxidants within the cell is crucial: it has to be at the mitochondria. Every other strategy tried to date does nothing by comparison, or actually causes harm by interfering in the use of reactive oxygen species as signaling molecules.

The article references Protandim as "shown in a published human clinical trial to raise S.O.D. levels." This is in fact a dubious interpretation of the trial data, pushed by the manufacturer for all the obvious reasons. Protandim is an excellent example of the sort of marketing and unproven product that makes it hard for people to talk seriously about developing longevity therapies. By way of comparison, in the Skulachev laboratory, Russian researchers have demonstrated a 20–30% extension of life in mice using their engineered antioxidant compound. I think it's safe to say that if your antioxidant therapy doesn't extend healthy life in animal studies, then you don't really have anything of interest.

So overall not a good article at all; it's pretty much disinformation. h+ Magazine can do better than this.

Selling Fasting

Fight Aging!, 15 October 2016

One of the more recent innovations in calorie restriction research has nothing to do with the science, and everything to do with figuring out how to pull more funding into the field. There is never enough funding for research in any field: going by how funds flow through our societies, it is easy to say that to a first approximation no-one really cares about progress in medicine. Bread and circuses, yes. Better technologies, better understanding of biology, and less disease, no. There is also a large difference between the funds available for non-commercial research versus money available and interested in investment in for-profit ventures. The latter is at least ten times the former, and much more easily arranged as well. Writing grants and raising philanthropic funding is a considerably harder job than pitching angels and venture firms; more effort for fewer dollars at the end of the day. But without the funding for non-profit research initiatives, there will be no new technologies ready to be carried forward in for-profit companies. It is one of the great frustrations of patient advocacy to know that the owners of countless millions of dollars are sitting on their hands, waiting for viable biotech companies, while the important research projects that will generate those companies struggle to raise hundreds of thousands to sustain shoestring budgets.

Calorie restriction is a particular challenge in this context. It is a lifestyle choice, not a drug or an antibody or something else that the medical industry understands how to package, market, and sell. It is nothing more than eating sensibly and eating less. Anyone can choose to do it. It is free and straightforward and well-documented. Yet the effects on long-term health and aging in ordinary individuals are much larger than anything that can be generated by the presently available panoply of drugs and other interventions. The nature of calorie restriction means that there has been little to no for-profit investment aiming to better characterize its benefits. Rather, all that funding was directed towards mapping the biochemistry and haphazardly testing the established drug libraries to find something that triggered any of the same effects. The search for such calorie restriction mimetics is well documented elsewhere, so I won't dwell on that, beyond noting that the outcome of ten to fifteen years of work and a great deal of money is, so far, nothing of any practical use.

So to calorie restriction itself, and how to obtain for-profit funding for research into eating less, and eating less in an effective way. The innovators here are Valter Longo and colleagues, who have achieved the goal of pulling in for-profit funding on the backs of turning specific implementations of fasting and low-calorie diets into FDA-approved therapies, such as an adjuvant in cancer treatment. The magic of regulation means that companies can manufacture a medical diet on the basis of research, and then use the barriers set up via intellectual property and regulatory pronouncements to charge an inordinate amount for what is, basically, a little bit of food that anyone could throw together after reading the papers to obtain the target calories, protein, micronutrient levels, and so on. That in turn means that the principals of these companies are willing to pay for the supporting research. On the one hand it's a depressing example of the distorted priorities that emerge from regulation of medicine, on the other one feels a certain admiration for Longo et al for having successfully hacked the system to fund the useful results they have produced these past few years. Quantifying the degree to which fasting alters the immune system, and quantifying the degree to which low-calorie diets and fasting are effectively equivalent in altering metabolism, are both helpful new information for those who practice forms of calorie restriction and intermittent fasting. In any case, here is the less useful outcome from all of this, which is to say the medical diet. It comes across as a bad parody of itself, but that seems fairly true of most medical diet products.

The company behind it, L-Nutra, markets its product, ProLon, as a five-day-a-month, plant-based meal program built to convince the body it is fasting while the patient still eats — what they call a "Fasting Mimicking Diet," developed at the University of Southern California's Longevity Institute with backing from the National Institute on Aging. The company reports that in a trial of 100 participants across three monthly cycles, the diet produced statistically significant improvements across a full roster of standard health metrics: body fat, body weight, bone density, fasting glucose and insulin resistance, cholesterol and triglycerides, the aging marker IGF-1, C-reactive protein, and markers of stem and progenitor cell activity. The daily menu runs 770 to 1,100 calories.

Needless to say you can do all of this yourself, and whether or not you happen to have cancer at the time. It isn't hard to construct and follow a diet to a specific target of calories and nutrients: it just takes the willingness to do it. When presented with claims like these, it has to be said that it is more of a challenge than usual to remain optimistic that the first generation of rejuvenation therapies after the SENS model, such as senescent cell clearance, will be able to do without the ridiculous marketing language that characterizes present day efforts such as this one.

The Debate That Decides How Long You Live

Fight Aging!, 2 September 2008

Consider researchers who believe engineered longevity must be accomplished by gene engineering and changing the operation of metabolism to slow aging. In that worldview, any significant progress is far in the future, because the task is very complex indeed. Progress in the future is also largely irrelevant to those of us alive today, as slowing aging does next to nothing for people who are already age-damaged to the point of disease and frailty.

I consider it to be unfortunate that the bulk of the pro-longevity aging research camp is focused on an inefficient path forward that will in the end lead to lesser benefits. It is their belief that this is the only practical way ahead: a laborious slog towards complete understanding of aging and metabolism, followed by an even more complex navigation through re-engineering that metabolism to age more slowly. The sheer scale and difficulty of that task is why many scientists feel that meaningful engineered longevity — more healthy years through science — is a long way away indeed.

Fortunately there is a fast boat in addition to the slow boat described above. It is likely to be easier and less costly to produce rejuvenation therapies than to produce a reliable and significant slowing of aging. A rejuvenation therapy doesn't require a whole new metabolism to be engineered, tested, and understood — it requires that we revert clearly identified changes to return to a metabolic model that we know works, as it's used by a few billion young people already. Those rejuvenation therapies will be far more effective than slowing aging in terms of additional years gained, since you can keep coming back to use them again and again. They will also help the aged, who are not helped at all by a therapy that merely slows aging.

Consider another manifestation of the "we can do no better than slow aging by metabolic manipulation" viewpoint, one that goes so far as to cast the alternative as arrogant. The argument, made in a paper whose very title charges this field with extreme arrogance, runs roughly as follows. Aging has many distinct causes, each an age-associated pathology studied in exhaustive detail by innumerable scientists the world over; to propose reversing all of them within a few decades is to claim success where that entire combined effort has so far fallen short. And since the finite life span of an animal is bound up with its anatomy and physiology — the product of an extremely long evolutionary history — the proposal amounts to undoing, in a generation, what took millions of years to arrive.

That view is wrong-headed, to say the least, but it is an output of the sort of worldview described above: a) that aging can only be slowed, b) that doing so requires the day to day operation of human biochemistry to be changed in non-trivial ways, and c) that this is a very tall order indeed for the medical science of the foreseeable future.

So, to point out the errors. Firstly, the causes of aging are not the pathologies of aging. Pathologies are end results — if dry rot is a cause, then failing wooden structural beams are the pathology. Today's prospective longevity engineers talk about causes, about the comparatively few types of biochemical damage that build up in our tissues to create many, many different forms of pathology.

If you want to tackle aging efficiently — and make no mistake, this whole debate is about efficiency — then pathology is the wrong place to start. If you work on patching up pathologies, then you are Canute against the tide. Present day gerontological medicine is largely playing the role of Canute because that has historically been the best medical science can do: throw a huge level of resources at treating the consequences of aging and gain little by it. That little was better than nothing for billions, but it was still little in the grand scheme of things.

We stand in the 21st century now, amidst the early years of a revolution in the capabilities of biotechnology, and scientists can move beyond the historical focus of medical technology on patching end results to work instead on prevention and repair of root causes.

I predict that this notion — that the field proposes to undo millions of years of evolution — will come back to haunt anyone who takes it as a knockdown argument. No-one wants to be on record as saying something as bone-headed as that. In the past few decades medical science has reversed any number of evolutionary consequences, some of which have billions of years behind them. As if the number of years a feature took to evolve has any bearing upon the development of medicine that acts upon it!

Biochemistry itself is hard to re-engineer for greater longevity through slower aging: every component is part of many different complex evolved systems, and you can't change one thing without profound side-effects rippling through every connected process.

But the modern longevity engineers, the heretical minority in the aging research community, are not taking that path forward — they simply want to restore the metabolism we had when young, not re-engineer a new one from scratch.

That is efficiency, and the nature of efficiency in longevity research is the most important debate within the life sciences today, for all that most people know nothing of it. The result of this debate will determine how long we all live in good health.

Interlude: 2015

Fight Aging!, 31 December 2015

I think that in years to come, we'll consider 2015 to be the point at which things really started to move for SENS rejuvenation research. It has to be said that medical research moves slowly at the best of times. It takes a good long run-up to show results, and it took a decade for SENS to grow from an idea and a few interested researchers to its present state of a foundation, the support of leaders in the research community, a loose network of research groups, and a few newly formed startup companies. It is perhaps an appropriate year for the unveiling of the Methuselah 300 monument, listing the donors who provided the initial funds and support to start this ball rolling.

Senescent cell clearance as a treatment for degenerative aging is having its breakout year. The Methuselah Foundation and SENS Research Foundation provided seed funding to Oisin Biotechnology for work in mice on a method of clearing senescent cells. Meanwhile another research group spent 2015 uncovering tissue specific drug candidates that tip senescent cells into the mode of programmed cell death called apoptosis, demonstrating this in mice with a couple of different drugs, producing results good enough to prove that meaningful health benefits result from a single treatment in older animals. This work should once and for all settle that SENS advocates have been right for the last decade, and the rest of the research community should have listened years ago.

It has also been a breakout year for gene therapy, but in a much bigger way. CRISPR has reached critical mass, and we are going to see an avalanche of gene therapies taking place in the years ahead both in trials and outside the regulatory system. A demonstration of this point was made by the BioViva CEO, who used medical tourism and the connections of a biotech startup company to undergo telomerase and follistatin gene therapies. Five years from now, it won't take any connections — just look up a reputable clinic outside the US and take a trip. If you look back at the progression of stem cell therapies since the turn of the century, that is exactly what is going to happen for the first gene therapies. Very few of the plausible candidates are what I'd call rejuvenation therapies at this point, more compensatory approaches that can spur additional stem cell activity or muscle growth or the like, but the growth in expertise in gene therapy in the field as a whole is a good thing for the future of SENS treatments that do require gene therapy.

Glucosepane cross-links are an important contribution to aging in humans — they are a part of the reason that skin and blood vessels lose their elasticity. The former is unfortunate, the latter ultimately fatal. For some years now the SENS Research Foundation has been funding efforts to develop the tools needed to work with glucosepane in cells and tissues, and this year a first success was published in a prestigious journal: a reliable method of synthesizing glucosepane as needed, a very important part of the toolkit.

Another lengthy SENS research program now blossoming is the use of allotopic expression of mitochondrial genes. The researchers originally funded by the Methuselah Foundation and SENS Research Foundation formed a company, GenSight, that is now well on its way to clinical application of this technology for inherited mitochondrial disease. That foundation of practice and experience will hopefully create a much better basis to finish up the work for all mitochondrial genes and the treatment of aging in the years ahead.

The long-running efforts by SENS researchers to find bacterial enzymes capable of breaking down some of the waste chemicals that form lipofuscin, a mix of metabolic wastes that clogs up lysosomes in old cells, have reached the point of commercial development. Candidate drugs have been licensed out to newly formed Human Rejuvenation Technologies, Inc. Now we wait and see how that goes, but in general you should consider these sorts of deals a way to bring in more money for later stage research — it's just less visible until it reaches its goals.

Even the cancer programs based on preventing the lengthening of telomeres at the SENS Research Foundation are getting more attention from publications, and are no longer alone in the research community. Other groups are striking out in their own attempts to suppress cancer by preventing cells from lengthening their telomeres.

Speaking of commercial development: this year's Rejuvenation Biotechnology conference was a success. The hand off from academia to industry doesn't just magically happen in any field, and longevity science is no exception to this rule. This conference series exists to build the relationships and awareness needed for a smooth transition of rejuvenation therapies from the laboratory to biotech startups and Big Pharma in the years ahead.

Amyloid clearance inside the Alzheimer's research community continues to be a long, slow, painful progress of advancement by small degrees. This is not the only sort of amyloid, however, and this year efforts to clear transthyretin amyloid — associated with heart disease in the old and heart failure in the very oldest individuals — met with success in a human trial. This is perhaps the most advanced of SENS repair therapies at this point in time, an actual honest to goodness narrow scope rejuvenation therapy with a successful trial behind it. Next up should be a sober assessment of how to jailbreak this advance out of the regulatory regime and into more general availability via medical tourism.

2015 also saw continued and louder debate on whether aging should be officially classified as a disease, a categorization fight with real financial and regulatory stakes, and I expect it to keep growing.

Crowdfunding by the longevity science community continues apace. We're still not as good at fundraising as our nearest cultural neighbors, the strong AI philosophy of development typified by the Machine Intelligence Research Institute, but a little friendly competition never hurt anyone. That branch of the strong AI community, like SENS and much of the radical life extension advocacy community, arose from the transhumanist communities of the past few decades, and many of the same people have interests on both sides of the fence. It seems reasonable to think that we can do just as well as MIRI and related organizations when it comes to pulling in funds for our goals in healthy longevity. This year the Fight Aging! matching fundraising held in collaboration with the SENS Research Foundation raised $250,000 for rejuvenation research, and an earlier crowdfunding project at Lifespan.io raised $45,000. We're improving year over year, but there's much further to go yet.

What about the large investments in our space? Calico continues to make deals and be a mystery to anyone other than the insiders, but increasingly looks like something halfway between Big Pharma and the NIH, which is to say irrelevant to any meaningful progress towards human longevity in the near term. It will be sad if it ends up another Ellison Medical Foundation, but that may well be where things are heading. The principals at Human Longevity Inc. are much more vocal in talking up their position, but are definitely not doing anything that is of great relevance to life extension — it's more a personalized medicine company with great PR, nothing exceptional. On a different and more positive note, philanthropist Peter Thiel has been talking more openly about his interest in longevity science this year: I think it is important that the people who are backing the more important ventures speak out in this way.

Off in the research community interested in slowing aging, the population of people without large amounts of funding in other words, advocacy for change is continuing. A trial of metformin to treat aging is being used to change the way in which the FDA views aging and therapies — at least I hope that's the intended goal here, as I have no expectation that this will move the needle on human health and life span. The Longevity Dividend advocates have, meanwhile, set down their scientific evidence and vision for more NIA funding for therapies to gently slow the aging process in a new book.

Parabiosis research, linking together the circulatory systems of an old and a young individual to identify factors that may influence or be influenced by aging, is proceeding apace. It is being used to identify various potential drug targets, mostly with the expectation of increasing stem cell activity in old individuals. The debate over the validity of previous discoveries, such as GDF-11, also continues.

The cryonics community provided a demonstration of memory retained in vitrified and restored nematode worms, a great piece of evidence to support the hypothesis that present cryonics practices are sufficient to preserve the fine structure of the human brain, and this was also the year James Bedford became the longest surviving human being to evade the final end of information-theoretic death.

PART V

Money, Law, and the Machine

Progress, creation, good lives and better medicine don't come from politicians. They do not create or build — all they do is shout, threaten, tear down, abuse the power we give them and lay down roadblocks on the way to the future.

— Fight Aging!, July 2004

This is where the book turns. Curing aging is not, finally, a science problem but a money-and-law problem — who funds the work, and whether the regulators will let the results reach a patient. Somewhere in the middle of these chapters the advocate stops writing about other people's companies and starts putting in his own money, and then building a company himself.

A Strategy for Advocacy

Fight Aging!, 25 May 2011

There are three parallel tracks along which the future development of longevity science must progress, and we'll reach the end goal only as rapidly as the slowest of the three tracks moves.

Science and Biotechnology

The most obvious of the tracks is that the technologies of rejuvenation must be developed. We already know what these technologies must take the form of, and initiatives like SENS describe the needed procedures in detail at the level of cells and molecular machinery.

But the biotechnologies of rejuvenation don't yet exist, and the many technology demonstrations of long-lived mice, flies, and worms in laboratories around the world are nothing but a warm-up for the main event. Even the amazing pace of progress in stem cell medicine and cancer research these days is just a toe across the starting line when it comes to true rejuvenation. A great deal of work lies ahead, for all that it is very clear just what that work must be.

Clinical Development

Taking the output of the scientific method and turning it into reliable, affordable, widely available technology is no less a challenge than scientific progress. It is fraught with risk, and the stakes are much higher: the cost of developing new science is small in comparison to the costs of building an industry. Moving from one technology demonstration and a few patients to a technology used by hundreds of thousands of patients around the world is a massive undertaking in risk, development, innovation, and cutthroat competition.

Today, impersonal bureaucracy is busy crushing this track: the FDA, notoriously, outright forbids commercial development of therapies for aging.

That the track of development and clinical translation of research is lagging so badly is what prompted me to the vision of the Vegas Group. That in turn led to the recent launch of Open Cures — an initiative to help speed commercial development of longevity science. There are plausible, cost-effective ways in which matters can be put right, making use of the existing institutions of medical tourism, overseas research and development, the internet, and the growing community of garage biotech and open biotech developers.

Persuasion

People strongly enough in favor of engineered human longevity to get up and do something about it, who have a fair layman's or better grasp of the science, and thus know enough to support research like SENS rather than fall for any of the nonsense put out by the "anti-aging" industry, probably number a few thousand. There are probably tens of thousands more of a similar mindset, but not motivated enough to contribute materially beyond conversation and hope.

You can change the world with ten thousand people who think the same way as you do, that much is true, but it will be hard and it will take a long, long time. In the end, you'll only succeed by convincing hundreds of thousands more to contribute their support.

It doesn't require all the world to agree with you. Half a percentage point of the world's population is hundreds of millions of people. Markets with hundreds of millions of customers are worth billions of dollars, even though they gain only a tiny, tiny fraction of the attention and expenditure of those people — and billions of dollars would be more than enough to fund the first emergence of true rejuvenation biotechnology.

But it is a long way from where we are, a core group of thousands, to where we want to be — a core group of millions. Much remains to be accomplished.

Money Doesn't Matter; Methodology Does

Fight Aging!, 10 September 2014

There is an art to writing press releases for large joint ventures, and one part of it involves setting out the largest number you can vaguely justify in terms of dollars that will be spent in the future. You can look at the recent joint announcement of the Calico/AbbVie collaboration on longevity science as a good example of the type: of the $1.5 billion touted, everything above the first $100–200 million is basically fuzzy money, a matter of conditional future outlays, a hopeful position statement made far in advance rather than any sort of real commitment. Large numbers are rolled out in this way because the declaration helps the companies involved: it produces free advertising in the media circus, aids in gaining political leverage for tax advantages, and so forth.

But still, that first $100 million is a large chunk of change in the aging research world. The annual budget of the well-known Buck Institute for Research on Aging is a little more than $30 million, and the National Institute on Aging (NIA) budget is $1.2 billion for 2014.

If we take the usual ballpark guesstimate of public funding as about a third of overall research in this field in the US, that gives some idea of the scale of things.

Here is an opinion that I've heard expressed of late: why bother with all the effort and grassroots fundraising and advocacy to fund the rejuvenation biotechnology work at the SENS Research Foundation now that Google is throwing hundreds of millions of dollars into the aging science ring? This sounds like plain old lazy thinking to me. Not all research expenditure is the same, and not everyone who talks about tackling aging is in fact performing useful work likely to have much of an impact on human life spans in the near future. Consider that the organizations coordinating SENS research on repair of the causes of aging have been bootstrapped on philanthropic donations for a decade now, in an environment in which ten times the amount Google is likely to spend on aging research over the next few years has been expended by the NIA alone — and that is each and every year. The question could just as well be "why bother with SENS when the NIA is spending a river of dollars?"

"But look at the Calico website, right, they say they are tackling aging. That is new and different." That too is something I have heard. But this is really no different from the messaging you'll find at the NIA, whose own mission statement talks at length of understanding the nature of aging and extending the healthy, active years of life for older adults — without ever committing itself to anything so specific as repairing the damage that causes it.

This story really isn't about dollars. It is about methodology. If everything could be solved by simply ensuring a large inflow of dollars, then the world would be a much simpler and possibly much better place as a result. But how those dollars are spent matters far more than the amount. For all the rhetoric and grand budget of the NIA, outside of some cancer and stem cell research, I would be extremely surprised to learn that more than a few million each year out of that vast flow of money actually funds any of the remaining

SENS-like lines of research capable of contributing meaningfully to extended healthy life spans. The same is true of the private research and development institutions. The mainstream simply isn't undertaking the needed work, and that is why we need a funded SENS research program.

SENS exists as a disruptive innovation in aging research: it is a conceptually novel way of approaching aging and its treatment, a better approach to using existing capabilities and knowledge to produce longer, healthier lives at the end of the research process. It focuses on repair and root causes, and will ultimately overtake the research community to replace the old way of doing things that is consuming vast sums to no good end. SENS will achieve this goal by producing meaningful results on measures of health and aging where other approaches do not, and at a fraction of the cost; that is the reasonable expectation based on the scientific underpinnings, to my eyes, and to the eyes of a significant minority in the research community. Disruption is a bootstrapping process, a start from nothing but an idea, followed by incremental growth, proof, and persuasion until everyone admits you were right all along and switches to do things your way. This happens constantly in the technology business, and also elsewhere in the sciences, albeit on a slower timeframe, since the issues at hand tend to be more complex.

The deployment of large sums of money in any industry is an extremely conservative business. It is very, very rare for large institutions to head off in new radical directions — or indeed to intentionally take any sort of similarly large risk. They follow and reinforce the mainstream. At this time in aging research the mainstream is still characterized by the NIA and Big Pharma approach to age-related disease: treat the named diseases after the damage is done, rather than repair the root-cause damage that produces them. The disruptive adoption of a SENS approach of prevention and rejuvenation of aging and age-related disease has not yet happened, for all that it is on the way.

So when you see the emergence of an organization like Calico, well-funded, and headed by establishment figures from the research mainstream, then the odds are good that the organization will prove to be a continuation of the present work of the mainstream. It will most likely start out as a Big Pharma operation trying to make age-slowing candidate drugs work — akin to more of the same failed, expensive work on sirtuins and other aspects of the calorie restriction response, and similar lines of investigation.

As a part of the mainstream, Calico, like all the other existing large entities funding research, can be disrupted in the future, however. They will turn to devote funds to new methodologies demonstrated to be far more effective and cost-effective than the existing very poor paths forward in drug development to slow aging. The best way to have these organizations devote significant funding to rejuvenation research after the SENS model is for the SENS Research Foundation and related groups to be funded well enough over the next few years to be able to prove their case: to make one or more of the detailed proposed treatments work, and show that the results are far better and far less costly than the present dominant approaches to aging. With money that won't be too challenging, but "with money" is the hard part.

That is why we raise funds for SENS research: to ensure that it is adopted as soon as possible by the mainstream, because the diverse evidence from the research community paints a very convincing picture that SENS is the best shot at actually defeating aging. It is always a matter of raising a few dollars in the middle of a river of money heading toward the old, far worse, mainstream ways of doing things.

Why Rich Sick People Don't Fund Cures

Fight Aging!, 13 May 2016

There are a lot of people who have both a medical condition and a lot of wealth — tens of millions of dollars or more. In this day and age, a fraction of that wealth is enough to produce a prototype treatment from scratch for many classes of condition, if you are willing to wait the decade or two that low-cost basic science takes to run its course. Alternatively, for a faster result in the five-year range, that much money is enough to take a couple of promising potential therapies with initial animal studies and move them to prototype status. Not all conditions are amenable to this sort of approach, but many are. When you have a prototype, you license it freely to maximize the odds that it will be picked up and improved upon, and meanwhile pay a reputable clinic in one of the less regulated portions of the world to set it up for your own use. This is all very possible for a wide range of medical conditions. Why is it that so few wealthy, sick people take this path?

In the longevity science community we tend to ponder a very narrow facet of this question, which is to ask why, with very few exceptions, the wealthy of the world are not funding rejuvenation research. They are all aging to death, just like the rest of us. Why are they walking off the cliff when they have a good shot at preventing that outcome? Yet the broader question is also of interest: not just aging research, but all medical research. I was pondering this after donating to the present crowdfunding initiative for DRACO, a universal basis for cheaply creating effective treatments for any and all viral infections, such as those that are poorly controlled and afflicting large numbers of people today. How many individuals are there with both resistant viral hepatitis and enough money to take DRACO to what its inventor considers the finish line of readiness for human trials? The cost of that is a few million dollars at this point. I can think of a couple of individuals from the last celebrity generation alone who are in this demographic. But of course it isn't happening; these people are not jumping in to make waves and build out a prototype therapy that could cure or control their infections. So it seems to me that perhaps our first problem with regard to funding rejuvenation research isn't in fact a matter of convincing the world that treating aging is a viable goal in medicine. That is a challenge, and has to be accomplished, but it isn't the first issue in line. That first problem is that next to no one with the wealth to have a fair shot at solving their own medical problems through funding research thinks that they can in fact achieve that goal.

We can debate as to just why this is the case. For example, firstly there is simple ignorance of the possibilities. Some people and their supporting networks don't have the framework of ideas that lights the way. I think it isn't unfair to say that most people don't have any great insight into medicine as a system that can be changed and improved. I certainly didn't for half of my life. Engaging with doctors and learning about a specific condition because you happen to have developed it may or may not provide that insight — it strongly depends on the individual. The state of medicine, and even the state of waiting for better medicine, can be taken as set in stone. You can be good enough at what you do to become very wealthy, and yet lack the ability or patience or drive — or that framework of ideas — to learn the science behind the medicine, see that this science can be influenced, and understand the economics and connections well enough to see how to influence it. That is a tall order for someone who has invested decades in the minutiae of their own business and profession: a hard right turn in life, and a significant investment in time and will.

Secondly, there is a poster child effect here. Consider Michael J. Fox as one example, someone who has given large sums to Parkinson's disease research over the past two decades. Unfortunately, this is a condition in which it will take a long time and enormous funding to establish effective treatments, as is the case for most neurodegenerative diseases at this time. Conditions for which this is true tend to get a lot of the press, since there is more work taking place, and also more philanthropy. And the span of Fox's philanthropy crosses from a time in which life science work was very expensive and time-consuming into the present, in which it is much cheaper and faster. Medical research is very much easier today than it was at the turn of the century: all of the tools are greatly improved, as is knowledge of cellular biochemistry. But people think of this, and similar cases, and see decades of expense and no resulting cure. Subtleties such as the considerable progress achieved in both understanding the condition and building a foundation for treatments yet to come are somewhat lost on the world at large.

Thirdly, it is enormously expensive to move from prototype therapy to clinical availability through the regulatory gauntlet — the same over-the-top compliance regime, well covered elsewhere in these pages, that drives cost through the roof and restricts meaningful clinical development to entrenched Big Pharma interests. Building the prototype itself, by contrast, is cheap enough to be crowdfunded by ordinary people like you and me.

This is why I point out the strategy of open licensing and medical tourism: build the prototype, give it away, and undergo the treatment yourself. The BioViva CEO was (probably) the first human to undergo a particular gene therapy this way, for a low six-figure cost or less — regulation and its tremendous costs are not needed to produce a treatment judged safe enough to risk, and that choice should be up to the individual in any case. Yet near all of the people with the money to do this from scratch, and a condition that might be treated, don't see things this way. Wealth doesn't magically grant knowledge or wisdom; they view medicine as an enormously expensive undertaking, far beyond their ability to influence at all.

Putting My Own Money In

Fight Aging!, 17 February 2016

I'm pleased to be able to say that Fight Aging! has participated in the recent funding round for Oisin Biotechnologies, a senescent cell clearance startup founded by Gary Hudson. It was an unexpected opportunity to support this important line of SENS rejuvenation research, and will be my principal material contribution to the cause this year. From the point of view of where the money goes, there is actually little to no difference between investing in an early stage startup and making charitable donations to a laboratory group. In both cases the money buys research: lab time, reagents, mice, and the efforts of scientists. The closer things come to a working prototype, the more likely that someone will launch a company.

I consider it to be just as important to support the development of nascent SENS companies in their early stages as it is to fund the foundational work required prior to the point at which founding a startup becomes practical. An important part of the future of the rejuvenation biotechnology field is to create a virtuous cycle in which an ecosystem of growing companies feeds new funding back into fundamental research. The ideal situation for such a company is the one for Oisin Biotechnologies, in which the people and organizations with the largest ownership stakes and the earliest investment are all SENS insiders who are going to pour any realized gains back into research in one way or another. As for all startups in biotechnology, these are long bets, and there must be many of them in order to catch the few that spark into lasting flame. Most will fail, leaving only their research results, and the ones that succeed may take five years or so to get to the point at which funds can meaningfully flow back to research.

Ah, but ... all it takes is one SENS startup to do well enough, and, provided it is run by the right people, it will sweep up and carry forward all of the rest of the SENS agenda. One thing to remember about SENS rejuvenation biotechnology is that it is very cheap in comparison to, say, traditional drug development. At this point finishing the SENS agenda to produce first generation therapies in mice capable of repairing all of the primary forms of cell and tissue damage that cause aging — say half a billion to a billion dollars at this point — is much less than the cost of developing a single small molecule drug in the Big Pharma world, say two billion or so. A startup company in this field that made the transition to look something like a mid-sized pharmaceutical entity, with a market capitalization of billions, could probably finish up prototyping SENS on its own over a decade. It wouldn't be on its own, of course. If nothing else, the current clinical development of senescent cell clearance therapies, coupled with life span studies in mice, is going to wake up the world on the topic of rejuvenation. That is another good reason to support Oisin Biotechnologies.

At this point let me take a brief diversion into the evils of the US Securities and Exchange Commission (SEC). How is it that I knew about and had the chance to invest in Oisin Biotechnologies and you didn't? Simply because I'm close enough to being an insider in this community to get an invite. The rules put in place on early stage investment essentially act to forbid what is called general solicitation:

an early stage startup company can't simply advertise for investors. The founders can't reach out to the community at large. Raising a round cannot be public. This is a great example of regulatory capture at work. The rules, ostensibly to protect people from themselves — as heaven forbid anyone actually be trusted to make their own assessments of risk in this world — are absolutely and definitely shaped over the years for far less altruistic reasons. The goal is to restrict the opportunity to invest in high-risk, high-reward early stage companies to established networks of professionals, to build barriers and keep out anyone not on the inside.

This is changing, however. The advent of Kickstarter and its competitors has meant that suddenly a whole range of companies could bypass the whole idea of early stage investment in favor of mass preordering as a source of early funding. That works really well for manufacturing and creative efforts with a fairly short time frame. It is obviously much less useful for biotechnology and medical development. The SEC, for reasons that may have to do with the basic bureaucratic urge to control everything, or the interest of various parties in building new opportunities for regulatory capture, has altered its rules on early stage funding to permit general solicitation in a crowdfunding-like manner. Though of course, this being the SEC, it is legalistic, top-heavy, and people are still quite restricted in what they can invest. However, the basic point is that the investment process can be open and public, and in such a case anyone can invest. The new rules go into effect in the middle of 2016, and it remains to be seen how much of a mess or an opportunity it will be.

Mess or not, there is the potential to do something with this in our community. We are, modesty aside, pretty good at putting together and supporting modestly sized fundraisers for SENS research. If we can raise a quarter of a million in charitable donations for research, as happened last year, then I don't see it as beyond the pale that we could raise that much to crowdfund the founding of a future rejuvenation biotechnology company. Perhaps a glucosepane clearance venture, when that research gets to the point of a drug candidate, for example. Will this or something similar come to pass? Perhaps — it's at least possible.

2016 promises to be as active as 2015 for SENS charitable fundraising, based on what I know is coming up already. The Major Mouse Testing Program will be running a crowdfunding effort in the months ahead, and I think at least one other SENS-relevant group may do the same. When it comes to this year's main SENS fundraising in the last quarter of the year, however, I can't lead in the same way as I've done in past years by putting money on the table and telling the world to match it — what might have been those funds went to Oisin Biotechnologies and senescent cell research this year, an opportunity I could scarce turn down. Nonetheless, I believe we have plenty of time left in which to organize something interesting and useful, and I will still be the cheerleader to match the SENS Research Foundation's leadership when it comes to running the fundraiser.

The Off-Label Manifesto

Fight Aging!, 13 January 2023

What follows is a proposal for the best way forward, at the time of writing, to accelerate progress toward the development of diverse, effective rejuvenation therapies. The key is to use philanthropic funding to, first, prove efficacy in low-cost clinical trials, and second, market that data to ensure physician adoption of the first working rejuvenation therapies.

Executive Summary

1. Aging is by far the greatest cause of human morbidity and mortality.
2. Rejuvenation therapies that will greatly reduce unnecessary late life suffering and death are under development, but slowly, and with limited funding.
3. Accelerating the development of rejuvenation therapies is an important goal that, if achieved, will improve quality of life and save many lives in the years ahead.
4. At present there is only limited public support for the goal of human rejuvenation.

5. Faster progress will follow greater public and institutional support: more funding, more research programs, more biotech companies.
6. A few low-cost rejuvenation therapies exist today, already approved by US regulators for other uses. They can in principle be prescribed off-label in the US, but are not yet widely used.
7. Broad public and institutional support to accelerate the development of rejuvenation therapies will emerge following the widespread use of at least a few rejuvenation therapies.
8. Widespread use of the first rejuvenation therapies requires physician adoption of those therapies, as patients largely follow the options presented by physicians.
9. Physician adoption of an off-label rejuvenation therapy requires a convincing presentation of safety and efficacy in the treatment of common age-related conditions. This can be provided by favorable results from low-cost, few-hundred-patient clinical trials conducted by a reputable organization, and then marketed to physicians and physician organizations.
10. The cost of clinical trials to produce robust, trustworthy data for an existing therapy, already approved by regulators, can be a fraction of that required for clinical trials aimed at regulatory approval of a new treatment.
11. Given this funding, a number of existing organizations are well placed to conduct these trials and publicize the results. All that is needed is the will to act.

Aging Is the Greatest Cause of Human Morbidity and Mortality

Age-related disease is by far the greatest cause of human morbidity and mortality, and it traces back to the same handful of damage mechanisms covered throughout this book: senescent cells, cross-links, amyloids, DNA damage, and the rest.

The First Rejuvenation Therapies Exist

Historically, treatments for age-related disease have focused on the symptoms rather than the causes, and thus comparatively little headway has been made towards prevention and reversal — towards actual rejuvenation. Even though this remains true today, it is nonetheless the case that the first rejuvenation therapies exist, each capable of reversing to some degree a single contribution to degenerative aging.

For example, the senolytic combination of dasatinib and quercetin has been shown in human clinical trials to significantly reduce the burden of senescent cells in aged tissues following a single course of treatment. Senescent cells grow in number with age, and their inflammatory secretions are disruptive of tissue structure and function, contributing to many of the common fatal age-related diseases. Dasatinib is an FDA-approved drug that can be prescribed off-label at low cost, while quercetin is a supplement, readily available to all at an even lower cost.

As a second example, fecal microbiota transplant (FMT) from young to old individuals has been demonstrated in animal studies to produce a lasting reset of the aged gut microbiome and improved health. Restoring the balance of microbial populations in the intestine to a youthful state reduces the harmful effects of an aged gut microbiome on long-term health, including raised inflammation, increased generation of toxic metabolites, and reduced generation of beneficial metabolites. One implementation of this treatment was recently approved by the FDA for treatment of C. difficile infection. This makes it available off-label, in principle at least, and gives support to the community of practitioners that has for years been conducting FMT procedures on an ad hoc basis to treat various forms of dysbiosis.

But They Are Not Widely Adopted

Despite the existence of these rejuvenation therapies, and evidence for their benefits in the context of aging, the two treatments noted above are neither widely used nor widely known. The broader public still sees aging as largely set in stone, and treatment of aging as the domain of frauds and snake oil salesmen. The public does not realize that a small number of low-cost medical therapies can, with a single program of treatment, produce a lasting improvement to late life health by addressing one of the contributing causes of aging.

It must be noted that clearance of senescent cells and adjustment of the gut microbiome, while supported by evidence to show them to be individually beneficial, are not on their own sufficient for comprehensive rejuvenation of the old. Other contributing causes of aging must also be addressed — the damage categories covered elsewhere in this book. A faction within the biotech industry is working on new classes of rejuvenation therapy, and the funding for this effort has risen to a few billion dollars, spread unevenly across a few score companies. This sounds like a lot, but it is an insignificant fraction of the funding devoted to more mainstream medical development; the end to end cost of developing a single therapy was reportedly more than $2 billion in 2013.

If we wish to see rapid progress towards the effective treatment of aging within our lifetimes, a dramatic reduction of age-related disease, an end to frailty and dementia, then initiatives aimed at the production of rejuvenation therapies must greatly expand in size and number, and do so sooner rather than later. At the large scale, public support and understanding is necessary for the biggest, most conservative institutions to choose to put their shoulders to the wheel and advance the field of rejuvenation research. That support and understanding is missing at present, in large part because the first rejuvenation therapies are not widely used, and therefore remain largely unknown and unappreciated.

The problem is circular: little use of existing rejuvenation therapies means little knowledge of them, and little knowledge in turn ensures little use.

A slow process of bootstrapping is the usual way forward when faced with such self-reinforcing roadblocks, years of slow and incremental growth to open up a new field. There are other, faster paths ahead, however, enabled by the cost-effective deployment of philanthropic funding.

To Ensure Adoption, Persuade Physicians to Prescribe

The best way to break the cycle described above is to focus on physician adoption, to ensure that physicians and clinicians become comfortable prescribing and providing the few presently available rejuvenation therapies. Most patients only become familiar with the medical technologies presented by their physicians. Ensuring widespread off-label use by the medical community of the few existing, easily implemented, low-cost rejuvenation therapies would lead to a sizable improvement in public knowledge and attitudes regarding human rejuvenation.

A world in which most physicians routinely prescribe beneficial rejuvenation therapies is also a world that alleviates real suffering and mortality among older patients right now — a good in itself, whatever else it accomplishes for public support down the line.

Clinical Trial Data Persuades Physicians

The best way to convince physicians that a given FDA-approved treatment can be used off-label to treat many diseases of aging, or aging itself, is for reputable organizations to publish favorable results from multiple human clinical trials that employ the treatment. Such trials do not have to be anywhere near as expensive as the formal clinical trials conducted to persuade the FDA to approve a new therapy, as they do not need to be encumbered by the full set of regulatory concerns. They only need to be well run, such that the data produced is of high quality, and run by organizations with a good reputation, to ensure that the message is heard.

Trials of the few existing rejuvenation therapies are taking place, but only to a limited degree, and very slowly. The Mayo Clinic was one of the first groups to run clinical trials using dasatinib and quercetin in kidney disease, pulmonary fibrosis, and Alzheimer's disease, for example. But only for these few age-related conditions, with a limited budget, and at a sedate pace over years. Data has been published for only two of these trials over the last five years. Much more than the work of this single organization is needed to move the needle when it comes to persuading physicians.

It is possible to run a well-managed clinical trial of a low-cost, FDA-approved therapy in as many as two hundred patients for under $1 million, provided that the primary goal is to produce robust data rather than to satisfy regulators. Consider the PEARL trial for rapamycin, for example, which was crowdfunded with under $200,000 in charitable donations. Trials for dasatinib and quercetin treatment or fecal microbiota transplantation need not be that much more expensive.

Existing Organizations Can Administer Clinical Trials

A number of reputable organizations are well positioned to undertake the task of trial administration: connections to potential principal investigators and clinics capable of doing the work, project management as it progresses, and analysis and presentation of the results. This is far less onerous for the type of clinical trial envisaged here than for formal trials conducted for regulatory approval of new therapies.

The non-profit Lifespan.io runs crowdfunding campaigns to support research goals, and has collaborated with AgelessRx to fund and organize the PEARL clinical trial for rapamycin. The non-profit Forever Healthy Foundation performs research analysis and holds conferences, is connected to the Kizoo Technology Ventures fund that invests in biotech companies, and could make the transition to organizing small trials. The LongevityTech investment fund is constructing a network of clinics in a number of different locations for the express purpose of running first-in-human clinical trials for new therapies relevant to aging.

Outreach to Physicians

Given favorable data and reputable trial-running organizations, outreach to physicians is more or less a solved problem. It only requires funding and effort. Organizations that market therapies to physicians and clinical practices exist, while established physician networks and conferences can act as channels for broader discussion and outreach.

The first rejuvenation therapies worthy of the name already exist, but are barely used. Proving their efficacy and driving physician adoption, through rapid, low-cost clinical trials and the marketing of favorable results, is the fastest route from here to a world that takes rejuvenation seriously.

Plausible, Possible, Expensive, Prohibited

Fight Aging!, 3 May 2012

As an idle line of thought, what could you have done to yourself today in the field of cutting edge medicine and biotechnology at a moderate to high cost, setting aside the oppressive prohibition of medical regulation? Absent entities like the FDA — and a million other government employee busybodies who itch to regiment and enforce every aspect of our lives — it would be perfectly possible to get out there and solicit deals with researchers and clinics to try new things: to take your own estimate of risk and benefit, rather than being forced to wait for years or decades longer for medical technologies that might in the end be blocked entirely thanks to regulatory costs.

But what could you do today in a world that was more free, and with enough money to pay for a major medical procedure? Here are a few examples with varying risk-reward profiles, pulled from the air:

- Have your aging immune system wiped out with chemotherapy and replaced from your stem cells. Your wager here would be that undergoing chemotherapy (not a wonderful experience under the best of circumstances) will cause you less harm in the long term than keeping your original, increasingly misconfigured immune system. Alternately, you could wait a decade for targeted cell-killer therapies demonstrated in mice to become a practical concern in humans.

- Undergo any one of a number of potential enhancing gene therapies. For example, why not pay your way into possessing a myostatin mutation? That boosts muscle mass, increases resistance to a range of age-related conditions, and otherwise seems to be beneficial all-round in mammals.
- Purchase stem cell infusions of the sort that seem to be at least modestly helpful for any number of degenerative conditions — a better option than traditional pharmaceutical medicines. But of course you can't do that in the US. You'd have to head overseas as a medical tourist to become a customer of the more reliable clinics in Asia or the Middle East.
- Decide in your healthy old age that the possible benefits outweigh the risks for infusion-based bisphosphonate therapy. Of course you can't obtain that legally as a healthy person — those regulators again, deciding that they know better, and anyone who disagrees with them will ultimately wind up in jail.
- Choose to end your own long-lived life in a safe and painless way at the time of your choosing, while attended by cryonics professionals who can provide an immediate and expert preservation — offering absolutely the best chance of later restoration with minimal damage, while keeping the cost to a sensible minimum thanks to scheduling.

But of course any group that gathered in the US to try these things, or offer services, or make the process as safe and transparent as possible would quickly find themselves prosecuted and jailed. The land of the free long ago ceased to have much to do with liberty or personal freedom. Freedom is the freedom to take your own risks and pay the costs if you pull a bad card from the deck — and that freedom is exactly what drives progress. Take it away and what results is the regulatory stagnation you see in medicine today.

Let Patients Pay

Fight Aging!, 19 October 2017

There are a great many people willing to rain fire and brimstone upon the merest mention of clinical studies in which the patient pays the costs. They emerge whenever any group runs such a study, or whenever an initiative earnestly proposes greater adoption of this structure of trial for medical procedures. At root, there is a sizable fraction of the population that seems very hostile to the idea that patients be allowed to choose their own risks, and we see this hostility whenever some loosening or adjustment of regulation is proposed. Petty authoritarianism, yes, but to the degree that it shuts down responsible experimentation and progress in medicine, and slows the pace of development as a consequence, it is just as harmful as the real thing.

A primary complaint about the patient paid structure is that it is impractical to conduct blind studies or use placebo treatments in a control group, as people tend to want to get what they paid for. This greatly reduces the ability of a study to assess the presence or absence of effects that are small, intermittent, or vary widely between patients due to as-yet unknown factors. If you are looking for small and unreliable effects, and sadly this is pretty much the case for most modern medicine for the treatment of age-related disease, then this is a fair criticism. You should not be trying to use this sort of trial structure in such a scenario.

However, when it comes to rejuvenation therapies we are not interested in small and unreliable effects. Small and unreliable is the same as failure. The SENS rejuvenation research programs, and any other initiative to repair the causes of aging, is aiming for large and reliable effects. The expectation is certainly there for success in addressing any of the molecular damage at the root of aging to produce reliably beneficial outcomes in patients. But people are used to the marginal effectiveness of present day medicine, the quest for tiny, incremental, unreliable gains, and that steers their expectations regarding the appropriate way to proceed.

When it comes to placebo treatments, it is arguably the case that some types of patient paid rejuvenation therapy trial could be structured to include them. Consider senolytic treatments that clear senescent cells, for example. One could envisage a trial that involves two treatments two weeks apart. One is a placebo. After each treatment, metrics are assessed. In mice, beneficial changes following elimination of senescent cells occur quite rapidly, as the senescence-associated secretory phenotype is cleared out. There will be measures of health status in humans that can be checked a few days or a few weeks later. So in general, there are often ways to proceed with a placebo for short-term studies and short-term measures.

Yes, for longer-term patient paid studies the use of a placebo control group just isn't practical. It is perfectly viable to use the rest of the population as your control group. If a senolytic therapy cuts back all sorts of measures of metabolic age quite quickly, to a degree that can only be achieved via other methods over a long period of time, if at all, then it is hard to reject that out of hand just because there was no control group of study participants. Based on the results in mice to date, and what is known of cellular senescence in humans, one would expect the first drug or other therapy to replicate the same degree of cell clearance in humans to be so very evidently beneficial that attempts to run blind studies with a control group will be pointless — they will add little.

The response from the more conservative end of the scientific community following a patient paid study that shows significant benefits should be to pull in funds and interest to run their own more structured and careful studies to better quantify and improve upon the results. Publicity from the patient paid study will help them to do that, breaking through the reluctance that seems to characterize much of the industry. The patient paid structure is a way to bring in the necessary funding to carry out pilot studies in the absence of funding bodies within the existing institutions willing to do that. Given the history of neglect that senescent cell clearance suffered as a field before the landmark mouse studies proved that removing senescent cells could reverse aging, we should all be very skeptical that existing institutions will ever fund worthy projects in rejuvenation research if left to their own devices. They need to be kicked into action by outside efforts and outside funding, as they are otherwise happy to drift along performing marginal work.

If we want radical change to take place in the research community, if we want to see more work on human rejuvenation based on the SENS vision of repair therapies that can turn back aging, then patient paid studies are an important tool in the toolkit. Money doesn't grow on trees, and there are costs involved in getting the job done. Patient paid studies are one of the ways to overcome that hurdle.

Against the FDA

Fight Aging!, 16 April 2018

I am no great admirer of medical regulation as it presently exists in the wealthier parts of the world — the same burdensome system I've laid out before, one that costs multiples of what it should to bring medicine into the clinic, suppresses rather than encourages new medical technology, and tramples on the rights of patients to make their own informed choices.

In the past my prescription for a better future has been one in which medical tourism flourishes: the use of regulatory arbitrage to bring medicine to the clinic in a responsible way in places outside the US, born of a dysfunctional domestic system in which no party is incentivized to control costs and quoted prices bear little relation to actual ones. So let medical tourism grow as a form of pressure upon the existing regulatory system, because competition from outside is the only thing that has a hope of resulting in significant reform.

Any number of people work within the system on initiatives to make it better. There are non-profits and high net worth individuals and patient advocates all calling for reform of the FDA in one way or another. All are distinguished by having had very little positive effect over the years of their efforts. Indeed, over the last decade the cost of FDA approval has doubled, the time lengthened, and the rate at which new medical technologies are approved is stagnating. This has happened over the course of a period of enormous, staggering progress in biotechnology, in which costs of the underlying technologies have fallen dramatically. How can this be the case? Cynically, I would say this was also a period of great progress in connectivity and media exposure. Thus the tendency of bureaucrats to prioritize the minimization of bad press and accountability leads to ever greater demands, costs, and delays placed on regulatory applications, in search of a mythical, non-existent minimum level of risk that can only really be obtained by preventing new technologies from arriving at all.

Medical tourism, however, has challenges, primarily that it remains a highly disorganized market. Few of the people who might benefit significantly in fact do so. It requires a fair amount of research and determination; there is no easy on-ramp. I think that it will require universally desired enhancement therapies, such as follistatin or myostatin gene therapies for muscle growth, to obtain a large enough number of potential customers to generate sufficient organization in this marketplace to build that on-ramp. The number of people with severe illnesses is simply too small for the self-organization of entrepreneurship and venture funding to snowball in medical tourism — if it was going to happen, it would have happened already.

Here is another question: how do new therapies become available via medical tourism? Just as medical tourism is disorganized, so too there are large challenges facing company founders who want to responsibly develop and offer a therapy outside the established mainstream regulatory systems. A clear path forward to this goal has not been built; there is no roadmap. It is the jungle of uncertainty, and investors fear uncertainty. The founders of a company heading in this direction, such as BioViva Sciences and a few others, are faced with not just the challenge of building a therapy that is safe and reliable, but also building the entire infrastructure — the relationships, the legal understandings, the third party validation of safety and effectiveness — to bring that therapy into a clinic. Then they must also survive the bombardment of negative press from organizations invested in the status quo, hostile to the idea that proving safety and responsibility really doesn't require more than a fraction of the costs that the FDA imposes. Further, the founders must successfully woo investors who are very used to the present beaten path and nervous regarding anything new.

Taken together, this is a tall order. It is a task suited to a non-profit, or more accurately a dedicated organization, or a distributed bootstrapping process of cultural change driven by many such groups. The prize is well worth it: an acceleration of progress towards many new classes of therapy. As is so often the case, it is far easier to articulate the change we'd like to see than to make it happen, however.

So if I started a medical biotech company, I'd have to say that I would take it through the current regulatory system, while (a) being an honest player by the rules of that system, because that is the best way through the gauntlet, and (b) continuing to hold the view that the whole thing should be replaced with a far less self-serving, onerous, expensive, and terrible option. Unfortunately, the standard regulatory path is the only fairly reliable way to put a therapy in the hands of large numbers of patients. The other, ethically far better paths, such as that pursued by BioViva, have the unfortunate outcome of great uncertainty in whether patients will ever be able to use the treatment in large numbers. I'm sure I'm not alone in having come to this position.

The change the world needs here is clear: a far more organized industry of medical tourism, and an infrastructure for responsible medical development and validation of therapies, outside the established regulatory systems, that is generally accepted by investors. These challenges are easy to state, but resistant to any sort of easy solution that can do better than reaching a fraction of the patients who might benefit. At present any one company might succeed against the odds in obtaining enough funding and support to set up an offshore clinic — but how many patients can they reach? Small heroic battles, one company at a time, will not be enough. Until greater change is achieved, near everyone who starts the process of medical development will likely look at the landscape and make the rational short-term choice to pass through the existing regulatory gauntlet.

Dismantling Efficacy Review

Fight Aging!, 11 March 2025

Insofar as I have an opinion on politics, I'm against it. Generally it runs in its own noisy part of the world, involves a lot of unnecessary angst and drama, and at the end of the day has little to no influence on my day to day life. It is best ignored. I'll talk about it a little today because it seems likely that some upheaval lies ahead for the regulation of medical development in the US, and by extension that portion of the rest of the world that has outsourced much of its medical regulation to the US. The Department of Government Efficiency and related factions have yet to turn their eyes in earnest to the FDA, but they seem likely to do so.

One old idea that appears ripe for revival is to have the FDA cease to require proof of efficacy for new drugs, and only enforce proof of safety before permitting use in the market. By extension, this also implies preventing the FDA from attempts to sabotage the already permitted off-label use of existing approved therapies for new indications. The various payers in the system, everything from medical insurers to individuals, could then make their own decisions as to the proof and standards of efficacy they would like to see upheld before paying for a treatment. It would go back to being a market with a plurality of opinions on any given drug and use case in question, rather than having to accept the dictates of one set of bureaucrats, one size fits all.

One present manifestation of this ethos is the Right to Try law in Montana, which states that any drug that passes a phase 1 safety trial is open for use by any patient — though in practice obtaining that drug would be challenging for any novel therapeutic, requiring cooperation on the part of the developer that the leadership of that company might justifiably view as a risk to their ongoing relationship with the FDA. However, one could envisage a world in which this Right to Try approach effectively becomes federal law in the US, because that is a fairly simple change from the point of view of the executive branch: just alter the way the FDA behaves, and all clinical trial activity beyond phase 1 is optional. At that point, the incentives on developers change considerably, and the complexity thereafter is left to evolve in the market of consumers and payers.

It seems likely that the immediate response of the largest insurance companies and other payers would be along the lines of "great, but keep on running phase 2 and phase 3 clinical trials that demonstrate efficacy if you want us to pay for your medicines." I'd expect this to result in only short-term chaos for the network of investments, valuations, and stock prices that depend upon medicine being very expensive to bring to the clinic, and thus only short-term opposition from large investors, pharma entities, and other vested interests with political influence and lobbying capital to spend. After matters settle down to be more or less a continuation of the status quo, at least at first, greater freedom will emerge over time: payers will selectively defect from the consensus regarding standards for efficacy, where they feel it is justified, biotech and pharma companies will gain a much greater freedom to bring new medicine to the market earlier and at a lower cost than would otherwise be the case, and patients would gain the choice of earlier access. This is already happening via medical tourism for a few; the domestic option would broaden access enormously.

Another possible path forward is to make the US clinical trial ecosystem look a lot more like that in Australia. This would require more extensive changes to the regulatory system, which may or may not meet with opposition. In Australia there is no central government body akin to the FDA with the role of assessing and approving every phase 1 safety trial. Instead a competing market of specialized clinical sites and institutional review boards exists to assess and approve proposed new drugs and clinical trials, balancing their incentives to receive business versus their incentives to minimize harm to patients. Since something like 10% of all phase 1 trials worldwide are conducted in Australia, at something like half the cost and a fraction of the time required for approval in the US, it is a system that appears to work fairly well. Setting this up in the US would require dismantling regulations applying to institutional review boards and removing the FDA function of up-front vetting and approving of trials. As before, this change would likely produce short-term chaos and scrambling, but if the end goal remained to work towards the same present format of data and reporting that results from clinical trials, then the existing marketplace of contract research organizations (CROs) that run trials and package data would adapt, form standards, and those standards would likely look very much like the present status quo, at least at first. Moving forward, competition and freedom to choose would allow a greater plurality of options to emerge.

A harder and more disruptive change would be to alter the way in which the production of drugs is regulated in order to reduce the costs of compliance. Regulating manufacture under the banner of Good Manufacturing Practice (GMP) is a sizable chunk of the work conducted by the FDA, and conforming to FDA requirements on manufacturing processes is certainly a very large slice of the cost and effort required to bring a drug to the clinic. The guidelines put out by the FDA are standard and vague. Every single class of drug has accumulated over years and decades a deep and very detailed culture and tradition of non-public material and experience among consultants and regulators to describe the precise details of an acceptable interpretation of those vague standards. This cannot be discovered without engaging with regulators and consultants at considerable expense. Further, the requirements for GMP manufacture of any given drug class tend to expand over time, as efforts to remove any novel assay or other addition that regulators and consultants have seen used widely will be opposed.

Trying to change GMP requirements runs into a range of problems: firstly, there will be a great deal of political opposition from those who hold the unreasonable fear that any change to GMP regulation will make drugs unsafe; secondly, there will be a great deal of political opposition from large organizations that use the high cost of compliance to reduce competition, the usual problem of regulatory capture; thirdly, the specific interpretations of compliance that have become so very costly have very little to do with the published regulations, vary widely by drug type, are very complicated, and are near entirely documented privately, making them hard to target; fourthly, many smaller countries reference US GMP standards and rely upon them to inform their own medical regulation, essentially outsourcing that function.

To return to Australia, for example, the consensus that has emerged among institutional review boards, clinical trial running CROs, and clinical sites, is that the drug used in a phase 1 clinical trial should be manufactured to GMP-like quality. What "GMP-like" means in practice is that the FDA has accepted the proposed technical/scientific details of the manufacturing process and quality control assays, the drug batch is manufactured using that process and those assays, but the manufacture is not conducted with the very expensive addition of the full audit trail, checking in triplicate, form-filling, checkboxes, and validation at every step that is required for a process to be GMP grade. In drug manufacture circles this is called an engineering batch, and in present practice is the last full scale non-GMP batch that successfully tests all of the manufacturing processes and passes all of the quality assurance tests. The engineering batch is usually used for toxicity studies in animals that are conducted as a part of submitting an IND proposal to the FDA — except that in Australia one can also use the engineering batch for a phase 1 safety trial in human volunteers.

If engineering batches are safe enough for humans in Australia, why not everywhere — why not make it the minimum regulated safety requirement for manufactured batches at any stage of approval or commercial use? This would reduce manufacture cost by 50% for many drug classes. The batches would still have to pass quality testing that has been reviewed by regulators, institutional review boards, and others. The answer to the question "why not?" is probably that (a) this would be a hard change to make, politically, and (b) while it is easy to state the change in a single sentence, actually wrangling into shape the vague regulations and hidden details of the present state of the art regarding compliance would be challenging.

But it seems likely that we shall see how this all turns out!

Fix Aging First, Politics Second

Fight Aging!, 26 June 2004

I have strong libertarian leanings. I value freedom, honesty, and a strong rule of law. In an ideal world, I'd be able to live in a thriving minarchy or anarcho-capitalist society. I think that the Austrian school of economics provides a far better view of the world than that offered by competing economic theories.

Western style democracies are about as far as you can get from a libertarian society and still have a place that's moderately pleasant, free and safe to live in — so long as you blend in and don't make enemies amongst the powerful. The only high profile modern alternatives — dictatorships — are invariably very much worse places to live, after all. But it is simply not true that we need high taxes, large government, and unaccountable officials: there are better ways of doing things.

In any case, how does this little sidebar fit in with healthy life extension? The answer is that in a world of large government, in which a good 35% of all medical research funding is provided by government grants, you can't ignore the system. It is an open question as to whether public funding for medicine speeds the rate at which real anti-aging and healthy life extension therapies are made available more than other aspects of a large government slow things down. For example, high taxes have a devastating effect on the economic engine that powers growth, commercialization and advancing technology.

People are making serious organized attempts to create more libertarian societies, but this has little direct bearing on the large scale advance of real anti-aging science over the next decade or two.

Hence, we come to pragmatism. Between the two biggest problems I see in life — a) that we are aging and will all suffer and die if nothing is done, and b) that society is far less free and honest than it could be — aging is clearly the problem to be dealt with first. There's a time limit attached to it, and I am very much a first things first type of person.

So I draw my lines in a pragmatic manner. At the moment, I support working within the system to the extent of protesting government restrictions on research (although a more libertarian position would be that any such interaction with government has the undesirable effect of legitimizing the very system you oppose). I'm not a big fan of the California stem cell research ballot initiative, but only because I think that too much (big government restriction and interference) bad is coming with the (research dollars) good on that one. Private projects like the Methuselah Mouse Prize, or any number of foundations and research groups are just fine in my book.

People should draw their lines where they please. I'm not a jealous libertarian, like some I could mention, and I'm certainly not telling you what to do. If you are comfortable with living in a Western democracy, more power to you; in a truly plural society you would be able to do so without forcing me to do so as well. But we don't yet have such a thing, and probably won't until new frontiers — space, virtual nations, ocean living, floating cities in the upper atmosphere, and so forth — are opened up through advancing technology.

My advice to you is to support the healthy life extension initiatives you are comfortable with. If we all make smart choices, we can help medical science to advance towards real anti-aging medicine in the face of opposition, government and otherwise.

Death and Taxes

Fight Aging!, 24 December 2006

Death and taxes, the age-old twosome. There are certain similarities between attitudes and possible approaches to the two; death by degenerative aging on the one hand and government theft backed by threat of force on the other.

The first similarity of note lies in behaviors of learned helplessness and rationalization of the unpleasant when faced with aging and taxation. Sure, people worm around and try to make their position as least uncomfortable as possible — but the majority are doing so within the rules laid down, believing that death by aging and taxation are both set in stone.

The second item of note is that by banding together we can dramatically change or remove both death by aging (through the advance of biotechnology) and taxation (through changing our society). The only way the world changes is through human action, whether that action is the discovery of science or the changing of minds. Neither death by aging nor taxation is set in stone, but changing their solid present existence for everyone will first require sufficient ongoing education to change the expectations and level of understanding for most people — to change them from supporters of the status quo, no matter how bad, to supporters of a better future, and willing participants in bringing about that future.

On the one hand, we have the understanding of economics and expectations of government, and on the other hand we have the understanding of progress in biotechnology and the expectations of medical science. For most people, at this time, dearly-held beliefs and instinctive responses are widely divergent from reality.

These great tasks are very hard, but not impossible; distributed, co-operating groups of humans have attained equally challenging goals in the past. Societies and science can be made to change dramatically in half a lifetime. The development of a toolkit for the repair of age-related damage, and elimination of parasitic government and taxation, would lead to a society far less burdened by waste, both of lives and resources — a society in which growth, progress and individual lives are far greater than in the present time.

PART VI

Doing It Yourself

Everyone wants that silver bullet to be available now rather than tomorrow, and wants it badly enough to buy lead painted up to a nice sheen if that's all there is.

— Fight Aging!, March 2012

He does not only argue. Here he does the work on his own body — sourcing compounds, running single-person trials, taking senolytics and reporting the results, transplanting a younger person's gut microbiome into his own — with the same engineer's care he brings to everything, and a candour about risk that most would not offer.

What Is Actually Available Today

Fight Aging!, 29 December 2015

At the present time it is technically possible to benefit from a small range of therapies that we can reasonably expect to at least modestly slow or reverse specific narrow aspects of aging. I'm not talking about supplements or calorie restriction mimetics here — we've already established that those are worthless next to exercise and calorie restriction themselves. What I have in mind here is instead the first gene therapies, stem cell transplants, and glimmerings of SENS-like repair therapies capable of removing some of the metabolic wastes associated with age-related diseases.

In order to actually undergo one of these new therapies, you would have to undertake some combination of the following: (a) spend money at early adopter levels, high in comparison to the cost a customer would pay for a final product years down the line, (b) network for connections to find access to the necessary services and other items, (c) persuade the small number of current developers to depart from their current practice of adhering to regulation and provide you access, (d) break (the unjust and largely horrible) laws related to provision of medical services, (e) travel to a less restrictive jurisdiction as a medical tourist, and (f) accept a fair degree of risk of failure — that even if everything else goes well, and all involved do their jobs, the present implementation of the treatment just doesn't work, or the present understanding of the science and data provides a false and inflated impression of what the treatment can achieve.

A good example of a combination of most of the above is the BioViva CEO's gene therapy from earlier this year: money and networking are the currency of a startup, and the existence of that startup was required to bring together all the necessary players to get the job done. That and medical tourism to a country where provision of the treatment is legal, as it would be everywhere in a just world. Sadly, in our world, there are entire branches of government in the wealthier nations staffed by those who, day in and day out, toil to block access to potential therapies for people who can make the educated decision to take the risk.

A few pointers, based on the work of recent years:

Gene Therapies

With the advent of CRISPR, the technical feasibility of gene therapy is now leaps and bounds ahead of where it was just a few years ago. The remaining challenge in delivery is obtaining reliable tissue coverage in adults; the results in terms of proportion of cells affected are very variable in animal studies, and understanding why these variations occur is an ongoing process. So you can undergo gene therapy and come out with too low a percentage of altered cells to make any meaningful difference.

That to one side, there are any number of gene therapies that are now technically feasible in humans. They span the spectrum of risk. In the most favorable camp are alterations already undertaken in animals for years, that researchers are practiced in, and which appear to be wholly beneficial, such as myostatin gene therapy to spur muscle growth and resist age-related muscle loss. In the least favorable camp there are alterations that could absolutely be set up quickly with CRISPR and a small lab, but have only been carried out in rodents a few times, and with limited long-term observation, such as adding extra lysosomal receptors to maintain youthful measures of liver function in old age via increased cellular garbage collection.

Stem Cell Therapies

If you are 60 years old in the US, with a typical level of wealth for that age group, and a typical level of creakiness in the joints, then why wouldn't you spend $10–20,000 on first generation stem cell therapies that have a good expectation value in terms of delivering relief from pain, control of inflammation, improved function, and so forth? The only reason people aren't doing this in droves is that it is still something that you have to know about, to pester the right doctors, to do some legwork on clinics and hospitals. For these very simple treatments you don't even have to leave the country anymore, since the FDA finally relented a few years back, but if you want to save money then heading to Canada or Mexico is very feasible.

If you are younger and your joints are not at that point of constantly reminding you of their damage, the cost-benefit analysis is much less clear. Will a healthy person in their 30s or 40s gain any meaningful benefit — short term or long term — from today's simple stem cell therapies? "No" seems like a very plausible answer to that question, there will be no useful data to help pin down the bounds of the possible for decades yet, and by that time it'll probably be irrelevant.

Enhancing Native Stem Cell Activities

Parabiosis research is uncovering signals in the blood that govern stem cell activities, such as GDF-11. Augmenting or reducing levels of these signals so as to spur greater stem cell repair and maintenance of tissues can be achieved via gene therapies of various sorts, as well as by targeted drugs. This is all still new enough that getting access to treatments would be a case of persuading one or more of a small circle of respected researchers to do this for you, and that just isn't going to happen outside of the context of licensing their intellectual property and funding their development process — things that look a lot like starting a company and building a technology. The risk here is also an unknown; will this spur cancer, or cause other interesting problems? There is too little work in rodents to even be certain of the present safety for laboratory animals.

Thymus Restoration

The loss of much of the thymus in early adulthood slows the supply of new immune cells to a trickle. Expanding that supply would be one way to reverse some of the age-related decline of the immune system. It is possible to grow thymic tissue from cells and transplant it, for example: both things have been done for humans, just not together in one patient. That is a very plausible near-term goal for a clinic outside the US that already has experience in cell therapies or tissue engineering. It is also possible to conduct gene therapy for FOXN1 to restore thymic activity — one of the many riskier options that CRISPR now makes possible should anyone be willing to do it for you.

Immune Cell Infusions

Researchers and many clinics are perfectly capable of generating immune cells to order, and in large numbers. It just isn't present practice to make a therapy out of this by delivering those immune cells to patients. It would be a matter of money and organization rather than new science to put together such a treatment given a clinic with the ability to carry out existing stem cell therapies. Provided regularly, such a therapy may well augment the aging immune system with large numbers of new immune cells capable of defending against pathogens and eliminating unwanted cells.

Clearance of Transthyretin Amyloid

Earlier this year, promising trial results were announced for clearance of transthyretin amyloid, a form of metabolic waste associated with mortality in the oldest of old people, as well as with heart disease in younger old people. The therapy is one of the first successful implementations of a SENS approach to aging, meaning repair of damage and clearance of wastes, but since it is being developed within the regulatory system it is being used to treat a specific age-related disease rather than as a therapy for a general form of damage that underpins many aspects of age-related degeneration.

The picture changes somewhat when this becomes a generally available clinical therapy, at which point things come back to medical tourism, but that is still years away for this particular approach to amyloid clearance given the slow pace of regulatory processes in medicine.

Bisphosphonates

Bisphosphonates were shown to grant a sizable five year increase in life expectancy in a study population of a little more than a hundred osteoarthritis patients. Obtaining and using drugs of this nature without a prescription — which you won't get unless you have osteoarthritis or one of the other conditions that bisphosphonates are used to treat — is of course going to be illegal in highly regulated nations like the US. The primary risk is that the data is incorrect, however; an effect of this size should not just appear out of the blue for a class of medicine that has been used for a long time. While researchers are investigating potential mechanisms that might explain these results, this seems a case of data that needs replication and confirmation rather than unquestioning acceptance.

Self-Experimentation, Responsibly

Fight Aging!, 27 April 2020

What follows is a walkthrough of setting up and running a self-experiment — a trial of one — to assess whether a ten week course of flagellin immunization will significantly and beneficially affect gut microbe populations. Flagellin is the protein making up a flagellum, the appendage bacteria use to move themselves around. As it happens, the presence of flagella correlates decently well with harmful gut microbes, and their absence correlates decently well with helpful gut microbes. In principle, provoking the immune system into greater efforts to chase down and destroy anything with a flagellum will better manage the microbial populations of the gut — populations that shift with age in ways that promote chronic inflammation and reduce the generation of beneficial metabolites.

Flagellin has been used in human clinical trials as an adjuvant to improve vaccination efficacy: it is recognized by the immune system, and helps produce a greater immune response. It was shown to have only minimal side effects in those trials. There is no human data for effects on the gut microbiome, but a very interesting paper describes the effects in mice: a shifting of the relative population sizes of gut microbes in a beneficial direction that will reduce chronic inflammation. It is unknown how long such an effect might last — whether rousing the immune system is a short-term process, or whether it will continue to better guard against unwanted microbes.

The purpose in setting out this outline is not to encourage people to immediately go and follow it. There are, for example, important caveats in the above-mentioned mouse study regarding links between flagellin immunity and human gut diseases. If you come away thinking that you should just jump in, and as soon as possible, then you have failed at reading comprehension. What follows is intended to illustrate how to think about self-experimentation in this field: set your constraints; identify likely approaches; do the research to fill in the necessary details; establish a plan of action; perhaps try out some parts of it in advance, such as the measurement portions, as they never quite work as expected; and most importantly identify whether or not the whole plan is worth actually trying, given all that is known of the risks involved. Ultimately that must be a personal choice.

Why Self-Experiment with Flagellin Immunization?

Gut microbes are an important influence on long-term health and aging. If forced to guess, they might be in the same ballpark as exercise. The first changes occur around age 35, reducing the generation of beneficial metabolites; later changes are more detrimental. There is at present little that people can do to reliably influence gut microbe populations beyond improving diet, there is considerable variability between individuals, and only a few services such as Viome offer assays to assess progress. Any sort of therapy that works for a majority of people without aggressive customization would be a step forward, and responsible self-experimentation can help to determine whether this is a viable path.

Caveats in More Detail

There are two areas of personal responsibility to consider here. Firstly, this involves injecting a bacterial protein that has little published data on human use, even where that data shows good safety in the short term. As the mouse study paper points out, human patients with inflammatory bowel disease exhibit greater immune reactivity to flagellin, and it is unclear whether this drives disease pathology or is a beneficial adaptation. That is a leap into the unknown. So is crossing the road, or indeed getting out of bed in the morning, but there are definitely different degrees of risk and comfort.

Secondly, obtaining flagellin in the manner described here is potentially illegal: not yet being a formally registered medical treatment, it falls into a nebulous area of regulatory and prosecutorial discretion as to which of the overly broad rules and laws might apply. In effect it is illegal if one of the representatives of the powers that be chooses to say it is illegal in any specific case, and there are few good guidelines as to how those decisions will be made. The clearest of the murky dividing lines is that it is legal to sell proteins not defined as a therapy for research use, but illegal to market and sell them for personal use in most circumstances. This is very selectively enforced, however, and reputable sellers simply declare that their products are not for personal use, while knowing full well that this is exactly what many of their customers are doing.

Choosing to purchase and use flagellin would therefore likely be a matter of civil disobedience, as is the case for anyone obtaining medicines or potential medicines outside the established national system of prescription and regulation. People are rarely prosecuted for doing so for personal use in the US — consider the legions of those who obtain medicines overseas for reasons of cost, despite the fact that doing so is illegal — but "rarely" is not "never." If you believe that the law is unjust, then by all means stand up against it, but accept that doing so carries the obvious risks of arrest, conviction, loss of livelihood, and all the other ways in which the cogs of modern society crush those who disagree with the powers that be.

Summarizing Flagellin Immunization

Flagellin is classified as a pathogen-associated molecular pattern, something that both innate and adaptive immune systems recognize and react to. The use of flagellin as an adjuvant, spurring a greater immune response to another attached protein or protein fragment, is an established field of development, largely because flagellin doesn't appear to be associated with mechanisms that might cause severe reactions. Human trials have taken place, and the safety data is good, at least over the short time frame in which safety is assessed.

Immunization with flagellin, as opposed to its use as an adjuvant, is a matter of injecting the protein on its own at an appropriate dose, to rouse the immune system to greater activity against anything connected to flagellin. Given that we mammals — mice and humans alike — are already sensitized to flagellin, it is interesting that boosting that reaction has a noticeable effect on gut microbes. That was only recently demonstrated in mice, and a larger literature on the topic has yet to be established.

Establishing Dosage

For small molecule drugs, it is possible to produce a starting point for human dosage given a mouse dose via standard equations, as shown in "A simple practice guide for dose conversion between animals and human." Unfortunately this doesn't apply to vaccines. In fact it is fair to say that there is no rigorous way to determine how vaccines scale from mice to people, since the systems involved are enormously complex: outcomes depend on immune system behavior, not on animal size, and modeling the immune system is a speculative activity at best. Many approved vaccines are probably using wildly suboptimal doses.

Insofar as there is any common wisdom, it is that similar doses in mice and humans appear to be a good starting point, based on those cases in which rigorous dose-finding has been carried out. In the case of flagellin, the mouse study employed a 10 μg dose injected weekly for 10 weeks. A 2017 human clinical trial used two monthly doses of 1 to 10 μg via intramuscular injection, and studies in non-human primates used doses of 1 to 10 μg.

Much larger doses have been given to animals without apparent toxicity, but the failure mode of immunization at too high a dose is intense inflammation or, worse, a cytokine storm that has the potential to be very harmful or even fatal. Whether such adverse consequences are triggered, and how likely they are, is species-dependent. There is no human data for higher doses than those mentioned above, so at the end of the day, the best starting point appears to be to stick with the 10 μg dose repeated weekly over 10 weeks via intramuscular injection, on the basis that this has been trialed in humans.

An Introduction to Injections

The relationship between different forms of injection, dosage, and effects is actually a complicated and surprisingly poorly mapped topic. There are four types of injection to consider, listed here in descending order of difficulty to carry out safely: intraperitoneal, through the stomach muscle into the abdominal body cavity, rare in human medicine but common in studies using small animals; intravenous, into a vein, which requires some practice to get right; intramuscular, into the muscle beneath the skin; and subcutaneous, into the lower levels of the skin.

The amount of fluid that can be easily injected varies by type. In humans, effectively unlimited amounts can be introduced via intraperitoneal or intravenous injection. The subcutaneous route is limited to something less than 1 ml, and intramuscular to 2–3 ml depending on location. These are all very fuzzy numbers, but the upper limits don't really matter for the purposes of injecting 10 μg of a protein: it can be dissolved in a very small amount of liquid, 0.5 ml or less.

Different injection routes can alter the character of the injected medicine — how much is required to gain a given effect, how long it takes to get into the system, and how fast it does so. A rare few medications cannot be injected subcutaneously because the metabolism of the skin will degrade them, while some are better given that way. Root through the literature looking for comparisons between performance and dosage across different injection types and you'll find a very ragged collection of examples showing few coherent rules. Some compounds show no discernible difference by route, some see altered peaks of concentration, some require higher doses subcutaneously, some require lower. Oil-based solutions can produce a very slow uptake of medication when injected into muscle or skin compared to an intravenous injection, while water-based solutions result in just as rapid an uptake into the bloodstream.

It seems sensible for a self-experimenter to use the much easier paths of subcutaneous and intramuscular injection, keeping the same dose as was established for intravenous injection. For most people, intravenous injections require a helper or a lot of painful practice; for subcutaneous and intramuscular injections, there is a market of autoinjection tools that removes many of the challenges. In the case of flagellin, it makes sense to stick with the human trial approach of intramuscular injections.

Considering Autoinjectors

Sticking a needle into one's own flesh is not an easy thing to do, and this is the rationale for the range of autoinjection systems developed by the medical community. They are most easily available for subcutaneous injections: spring-based devices that accept a standard needle and syringe, triggered by a button push. Intramuscular autoinjectors do exist, but unfortunately largely not in a general or easily available way — the needle-based intramuscular autoinjectors on the market are regulated devices that come preloaded with a particular medicine, and are not otherwise sold in a more generally useful way. There is no automation at all for intravenous injections; you are on your own there.

If intending to carry out subcutaneous injections, it is easy enough to order up a supply of disposable needles and syringes, an autoinjector device that accepts the standard needle and syringe arrangement, and other necessary items such as sterilization equipment from the sizable diabetes-focused marketplace. Such injections are relatively easy to carry out, a wide range of vendors sell the materials, and there is a lot of documentation, including videos, on how to do it. All of the equipment is cheap — though buying it will probably put you on a list in this era of the drug war, and there are many people out there doing it regardless.

The alternative is needle-free autoinjection, which can also solve the problem of missing general-purpose intramuscular autoinjectors. These systems use a thin, high-pressure fluid jet to punch medication through the skin, and are a growing area of development. They have numerous advantages over needles, but are more expensive, most can only manage subcutaneous injections, and all are limited in the amount of fluid they can inject compared to the traditional needle and syringe. Their biggest and most attractive advantage is in the name: not having to deal with needles in any way, shape, or form. There are a fair number of needle-free injectors on the market, but most are hard to obtain unless you happen to be a regulated medical facility placing a bulk order through the standard channels. From my survey of the marketplace, the two systems worth looking at are PharmaJet, which can be purchased in the US via intermediary suppliers, and Comfort-in, sold directly to consumers in most countries by an Australian group. So far as I can tell, PharmaJet is the only available needle-free system capable of intramuscular rather than subcutaneous injection.

PharmaJet is the better engineered and more expensive of the two, and its specialized 0.5 ml syringes are one-use only; loading fluid into them requires a vial and a vial adaptor, first loading the vial with fluid, then connecting it to the syringe via the adaptor to transfer the fluid. Comfort-in has a similar setup but is more flexible and more consumer-friendly overall, with a wider range of vial and other adaptors, and its syringes can in principle be reused given sterilization, though that isn't recommended. The instructions for both systems are extensive and include videos, and both are fairly easy to use. One caveat: needle-free systems produce a puncture that leaks injected material back out more readily than a needle does, so it's worth having a less absorbent plaster, such as one of the hydrocolloid dressings now widely available in stores, ready to apply immediately after injection.

If using the needle-free systems, vials of a specific type and size are necessary to fit the adaptors: a crimp-top vial, also called a serum vial, with a 13 mm (for PharmaJet and Comfort-in) or 20 mm (Comfort-in only) diameter open-top aluminum cap with a central hole to admit needles and adaptor spikes, and a rubber or rubber-like stopper thin enough in the center to let a needle or adaptor spike past. The cap is crimped on over the rubber seal to keep everything in place, which requires a crimping tool, and removing it requires another tool. You can either buy preassembled empty sterile vials plus a set of disposable needles and syringes to transfer liquid into them, or buy vials, stoppers, and aluminum caps separately, or as a kit, and assemble your own, plus a crimping tool of the matching size. Be careful when purchasing online, as vials are categorized by many different dimensions and descriptions tend to mix and match which ones they mention; for crimp-top vials, the dimension that needs to match the cap is the outer diameter of the mouth or crimp. It's usually best to buy a kit where possible, or if assembling from parts, to buy everything from the same company — Wheaton is a decent manufacturer whose equipment is widely resold, and their crimp-top 3 ml vials, snap-on rubber stoppers, and 13 mm open-top caps can be matched with a 13 mm crimping device and pliers to remove the caps later.

Preparing Flagellin for Injection

If using a needle-free injection system, you will likely be limited to injecting 0.5 ml amounts. The objective is therefore to obtain 10 μg of flagellin dissolved in 0.5 ml of phosphate buffered saline in each of ten sealed vials, ready for the injection system, with as little contamination as possible, stored in a freezer until ready to use. Stick to one dose per vial: repeated freeze-thaw cycles degrade a protein solution, which is what you'd be doing if all the doses lived in one larger vial.

When ordering flagellin, it will typically arrive in 10 μg to 100 μg vials. Empty 100 μg of flagellin into a single vial holding 5 ml of phosphate buffered saline, mix well, place that vial and ten empty vials into a vial rack, then use a pipette to transfer 0.5 ml into each empty vial, sealing and crimping each as you go before putting the whole set into the freezer.

Keeping hands, tools, vials, and surfaces clean and sterile matters throughout: wash everything carefully and wipe down surfaces with alcohol before and after use. Laboratories use autoclaves for steam sterilization; these are mostly expensive, but a range of smaller, cheaper options exists, and there are plenty of best-practices guides available online. This extends to the injection itself — even with needle-free systems, wipe the injection site down with alcohol first. It is all too easy to infect an injection site by skipping the precautions, with consequences that can be severe.

Not all flagellin is exactly the same molecule, nor is it produced the same way. The mouse study used flagellin from Salmonella typhimurium (there are numerous strains, but strain differences are probably irrelevant), typically produced via recombinant protein manufacture. Established companies that manufacture proteins to high standards of quality and safety generally won't ship to anyone but an established, validated lab business. The usual way for everyone else to obtain cost-effective supplies of this sort is to search Alibaba, though in this case not for flagellin directly, since it isn't listed as a stock item, but for protein synthesis companies willing to negotiate a price to synthesize it to order.

As noted at the outset, all of this — obtaining, shipping, and using any random protein for self-experimentation — is to some degree illegal, an act of civil disobedience carried out because the relevant laws are unjust, albeit very unevenly enforced. Many people regularly order pharmaceuticals from overseas, with and without prescriptions, and all of it is illegal; the usual worst outcome for individual users is intermittent confiscation of goods by customs (in the US, actually the FDA's job, not the customs authorities), with worse things reserved mostly for those reselling at scale or trafficking controlled substances. A mailbox capable of receiving signature-required packages from international shipping concerns such as DHL and FedEx will be needed — a business name and address is a good idea, and do not use a residential address.

Alibaba is the usual route to Chinese manufacturers for anyone without existing supplier relationships: a complicated site with its own culture and shorthand, forever shifting its payment systems and interfaces, but workable with some patience and a few good guides. Start by searching for protein synthesis companies — there are scores of biotech companies in China for any given specialty. Filter for small companies: larger ones tend to ignore individual purchasers after small amounts, and in any case demand the full paperwork of an established importer. Shop around, since prices vary widely and a low price isn't necessarily a red flag, and ask each manufacturer for mass spectra and liquid chromatography data if they have it. Ignore any stated minimum order size, however large; only deal with gold badge, trade assurance suppliers with several years of reputation and a decent response rate, and confirm the company has a real-world presence outside Alibaba, since storefronts, addresses, and account names don't always line up neatly even for entirely reliable businesses. Judgement improves with practice.

Once you have a few candidate suppliers, message them for a quote, providing the sequence and a reference to the paper in which it's described. Buy more than you think you'll need, packaged into multiple vials, since one vial will go toward validating the identity and quality of the batch. Payment will most likely be a wire transfer — what Alibaba calls telegraphic transfer (TT) — arranged against a pro-forma invoice (PI) and trusting the seller's honesty, which is fine for a supplier with a long-standing gold badge. Chinese manufacturers active on Alibaba are practiced at labeling shipments in ways that move smoothly through customs on both ends; proteins, unlike pharmaceuticals, are rarely confiscated even when the contents are declared honestly, so expect to pay customs duty if that's the case.

Proteins ship in a solid, freeze-dried (lyophilized) form, easily stored in a refrigerator short-term or a freezer long-term. Once mixed with liquid for injection, it has a much shorter life and should be kept frozen and used within a few months at most.

A protein may have been ordered, but that doesn't guarantee what turns up at the door is the right one, or free of impurities. Regular batch testing is good practice in any industry, distant supplier or not. The way to check is liquid chromatography and mass spectrometry against the standard data for a high-purity sample — or rather, to pay a small lab company to do that for you. Since splitting out microgram amounts or mailing protein in solution is more trouble than it's worth, just order an extra vial from the same batch and send that off for testing. Science Exchange is a solid way to find a lab for this: search for LC-MS services, pick a small company (large ones want all the registrations and paperwork dotted and crossed, and are a pain besides), and expect a little over $100 per sample. The request, bid, acceptance, and payment process runs through the Science Exchange site, with a discussion board for questions — most providers are happy to answer them. They'll want a description of the compound and its standard data sheet, which you can provide from the vendor or from NovoPro or other public sources. Ship the sample via a carrier like DHL, UPS, or FedEx, following whatever packaging guidance the lab provides: description and invoice on the package, order ID and sender/receiver referenced, sample containers clearly labeled, three layers of packing. Once the LC-MS process runs, the lab should provide a short summary of whether the compound is in fact the correct one, an estimated purity, and mass spectra you can compare against the vendor's own data.

Establishing Tests and Measures

There are a few options for testing before and after, the most compelling of which is the gut microbiome itself. Changes there, given a stable diet, should be indicative that something happened. But to assess outcomes beyond that, inflammatory markers should be tested at the very least; it might also be interesting to look at DNA methylation assessments of biological age, though it remains an open question what exactly these tests are measuring, and how useful they are to an individual.

The most direct measure of effectiveness is to assess gut microbe populations directly. There are lab services that do this, but few commercial direct-to-consumer options; one US-based service is Viome, though it doesn't present results in a useful tabular form and doesn't provide raw data. Nonetheless, it's a measurable endpoint, and gut microbe populations are fairly slow to change given a consistent diet, so large differences over a ten week course of flagellin injections would be interesting to observe.

For bloodwork, services such as WellnessFX let you order a blood test and have it carried out the next day by one of the widely available clinical service companies. Since inflammatory markers are the topic of interest, a service offering more individual tests rather than fixed packages might be the better choice — the Life Extension Foundation, for example, offers a wide range of tests including a good selection of inflammatory markers.

DNA methylation tests can be ordered from vendors such as Epimorphy / Zymo Research, though delivery takes a fair few weeks. From talking to people at these companies, the normal variability for repeat tests from the same sample runs to something like 1.7 years for the Zymo Research test; day-to-day or intraday variation between different samples from the same individual is more of an open question, though I'm told the results are quite consistent over measures separated by months. Either way, it's wise to keep everything as similar as possible when taking the test before and after treatment: time of day, recency of eating or exercise, recent diet, and so on.

Guesstimated Costs

The costs given here are rounded up for the sake of convenience, and in some cases are blurred median values standing in for the range of observed prices in the wild. The choice to use needles for subcutaneous injection is obviously much cheaper than exploring the world of needle-free injections and vial assembly.

- Business mailbox, such as from UPS: $250 / year
- Cytokine blood panel test from LEF: $300 / test
- MyDNAage kits: $310 / kit
- Viome kits: $150 / kit
- Miscellaneous equipment — spatulas, labels, vials, a vial rack, etc.: $60
- Small pack of 13 mm sterile serum vials: $35
- PharmaJet needle-free injection kit: $1020
- Comfort-in needle-free injection kit: $470
- Bulk 13 mm serum vial parts and capping tools: $750
- 100 μg of flagellin via Alibaba: $1000
- Shipping and LC-MS analysis of a sample: $200

Schedule for the Self-Experiment

Do you think you can reliably pipette fluid in 0.5 ml amounts between small vials, cap vials, connect adaptors, fill syringes, or carry out an injection without messing it up somewhere along the way? Perhaps you can — but it's a very good idea to practice first with saline solution rather than discovering that your manual dexterity or methods are lacking while handling an expensive protein. You will doubtless conclude that more tools, or different tools, are needed than you expected.

The process of discovery, reading around the topic, ordering materials, and validating an order of flagellin should be expected to take a couple of months. Once the decisions are made and the materials in hand, pick a start date. The schedule for the self-experiment is as follows:

- Day 0: perform the various tests — bloodwork, gut microbiome assessment, and so on.
- Days 1, 8, 15, 22, 29, 36, 43, 50, 57, and 64: intramuscular injection of 10 μg of flagellin.
- Day 65: repeat the tests.

Where to Publish?

If you run a self-experiment and keep the results to yourself, then you helped only yourself. The true benefit of rational, considered self-experimentation only begins to emerge when many members of a community share their data, to an extent that can help to inform formal trials and the direction of research and development. There are numerous communities of people who self-experiment with various compounds and interventions, with varying degrees of rigor; the LongeCity forums are a fair place to post the details and results of a personal trial, and equally, if you run your own website, why not there?

When publishing, include all of the measured data, the doses taken, duration of treatment, and age, weight, and gender. Fuzzing age to a less distinct five year range — late 40s, early 50s, and so on — is fine. If you wish to publish anonymously, it should be fairly safe to do so, as none of that data can be traced back to you without access to the bloodwork provider, and none of the usual suspects will be interested in going that far. Negative results are just as important as positive results. Publish whatever the outcome.

The Results, Three Years On

Fight Aging!, 3 April 2023

An earlier self-experiment in flagellin immunization set out to favorably adjust the aging gut microbiome by motivating the immune system to more aggressively destroy the problematic microbes that tend to grow in number with age. As assessed using the Viome service, it appeared to produce a sizable, lasting benefit to the quality of the gut microbiome in a basically healthy 50-ish individual.

There is animal data to show that fecal microbiota transplantation from a young individual to an old individual rejuvenates the gut microbiome as well — and a great deal more of it, produced in varied species, demonstrating that the improvement lasts for a lengthy period and even results in improved health and extended life span.

So the same individual later underwent a fecal microbiota transplant, using donor material from a healthy, athletic 20-year-old volunteer, and tracked the effects on the gut microbiome. Stool samples were tested beforehand, one month afterward, and six months afterward, all again via Viome. The flagellin immunization had been conducted in May 2020, the transplant at the end of August 2022.

While the flagellin intervention greatly reduced microbial diversity in the process of greatly improving other metrics, that diversity was restored by the transplant, and the restoration held through the six month mark. The transplant produced small gains in some of the other Viome metrics too, though not to the same degree as the flagellin immunization — likely because much of the potential scope for improvement had already been achieved. Interestingly, neither intervention did much for Metabolic Fitness. According to Viome, half the population falls into the 22–28 range for Metabolic Fitness on a scale of 0 to 100, which makes one wonder a little about the algorithm behind that particular number.

The protocol for conducting a fecal microbiota transplant at home is almost too simple to talk about, but there are a few points worth thinking through. The mechanics are straightforward: a fresh stool sample is provided by the donor and mixed with water, and a few fluid ounces of the result are delivered as an enema. The recipient then lies in a suitably sloped position, abdomen higher than chest, for 30 minutes or so, to encourage the enema fluid to flow as deeply as possible into the intestine. Repeat this two to three times, a few days to a week apart.

As to the points worth thinking through: when fecal microbiota transplantation is conducted in the clinic as a treatment for C. difficile infection, colonic cleansing and colonoscopy equipment may be used, but more importantly, donor stool samples are screened for potentially pathogenic microbes that an older individual may respond poorly to. This screening is wise for an older recipient, since the aged immune system is far less competent than a young one at suppressing undesirable microbial species in the gut — what is innocuous to a young person may be much less innocuous to an old person. Rather than going through the process of finding a willing, healthy volunteer and rolling the dice on potential issues, one can use services that will sell screened and characterized stool samples from young individuals, such as Human Microbes. This is recommended.

Rolling Your Own Senolytics

Fight Aging!, 12 January 2018

What follows walks through the process of setting up and running a self-experiment — a trial of one — of candidate senolytic drugs capable of removing some portion of the senescent cells that accumulate with age to cause aging and age-related disease. Metrics are assessed beforehand and afterwards in order to shed some light on whether or not it worked, in the sense of producing some degree of rejuvenation, turning back specific measures of age-related decline.

The outline here is optimized for simplicity, cost, and the ability to conduct the experiment without much outside assistance, rather than for maximal effectiveness. There are better candidate pharmaceuticals and better metrics than those settled on here, at least from the point of view of likely effectiveness, fewer side-effects, and relevance to the task at hand, but they require more work, more funds, more complicated logistics, and the assistance of laboratories and physicians. Given these self-imposed constraints, the outline here ends up focused on repurposed chemotherapeutics, which make up the majority of the current senolytic drug candidates. That in turn means that side-effects and related risks to health are an important consideration.

Why Self-Experiment with Senolytics?

Senolytic therapies are those that selectively destroy senescent cells. The build-up of senescent cells is one of the causes of aging. So obviously, one hope is to benefit personally from such a therapy sooner than would otherwise be the case, balancing that against incurring some unknown degree of risk of failure or harm. The first human trials, those that establish numbers for that risk, will take another few years to wind through to robust conclusions, and further years beyond that will be required for the medical community to become willing to prescribe senolytics generally. Further, those trials will almost all test only a single candidate therapy, and the evidence to date in mice suggests that different senolytics with different mechanisms are tissue-specific in their effects on senescent cells; multiple different compounds may be more effective than one, but that won't be discovered in the formal trial process. Lastly, well-run self-experimentation carried out by a number of people, where the results are published, can help to guide the direction of later, formal studies.

Caveats in More Detail

There are two areas of personal responsibility to consider here. Firstly, this involves taking chemotherapeutic pharmaceuticals with known side-effects. One should read the relevant papers on their effects, side-effects, and dosages, and make an individual decision on risk and comfort level based on that information. This is true of any pharmaceutical, whether or not approved for use. Do not trust other opinions you might read online: go to the primary sources, the scientific papers, and read those. Understand that where the primary data is sparse, it may well be wrong or incomplete in ways that will prove harmful. Also understand that older physiologies can be frail and vulnerable to the side-effects of specific chemotherapeutic pharmaceuticals in ways that do not occur in younger people and that are not well covered by the studies; pharmacokinetic studies necessary to establish side-effects and tolerances don't tend to be carried out in very old humans, and most cancer trials have participants that almost entirely fall into the 50–80 age range.

Secondly, obtaining these pharmaceuticals outside the prescription system is illegal — an act of civil disobedience against an unjust but very unevenly enforced law. People are rarely prosecuted for personal-use purchases, but "rarely" is not "never."

Lastly, senolytics is a fast-moving field. What follows will become outdated quite rapidly in its specifics regarding candidate pharmaceuticals, as new knowledge and new candidate therapies arrive on the scene. Nonetheless, the general outline should still be a useful basis for designing new self-experiments involving later and hopefully better compounds, as well as tests involving more logistical effort.

Choosing Senolytic Drug Candidates

The criteria for choosing senolytic drug candidates for the purposes of this outline are: it must be taken by mouth rather than through injection, as the logistics for assembling materials and carrying out injections are considerably more complicated; it must have shown senolytic effects in animal studies, not just in cell studies, as there are all too many failures to make the leap from cell to mouse in pharmaceutical development; and there must be enough human data to determine the effects and side-effects of doses used — the more, the better. Finding a list of senolytics and assessing them against these criteria involves research: dig through PubMed in search of senolytic studies and review articles, then follow chains of references to find other papers. Carefully check the magnitude and other details of the results claimed in animal studies, as some senolytics are better than others. Armed with the names of drug candidates, look up studies of dosage and effects for cancer and other trials in both mice and humans. Not all papers are open access; where they are not, taking advantage of the efforts of the copyright heretics of Sci-Hub is the best approach to obtain a copy.

At the present time, the criteria above narrow the field to a few repurposed chemotherapeutics, one of which was shown to synergize with the flavonoid quercetin. These are dasatinib in combination with quercetin, noting that the data shows it isn't worth trying either one on its own; navitoclax / ABT-263; and alvespimycin / 17-DMAG. Navitoclax, however, has side-effects that are common enough and severe enough to want to avoid it; it causes a loss of platelets in the blood to the point of producing noticeable medical consequences, and did so in about half of cancer trial participants.

There is also venetoclax / ABT-199 to consider, a modified form of navitoclax intended, and shown, to reduce the worst of its platelet-destroying side-effects, but lacking animal data for senolytic effects. Navitoclax and venetoclax are both BCL-2 family inhibitors, operating on roughly the same mechanism, and another member of this family of pharmaceuticals, ABT-737, has been shown to have senolytic effects. So while venetoclax lacks mouse data for senolytic effects, at first glance it makes some sense to include it in a test: the trade-off is a matter of losing some of the unpleasant side-effects and gaining uncertainty in whether or not the senolytic effects will carry over.

A handful of other candidates — A1331852, A1155463, piperlongumine, fisetin, FOXO4-DRI, and ABT-737 — are ruled out here for lacking published animal data, requiring injection, or both, though FOXO4-DRI in particular looks like it could be the best of the drug-like options so far discovered, if the animal data ever carries over into human tests.

Establishing Dosages

The only definitive way to establish a dosage for a pharmaceutical in order to achieve a given effect is to run a lot of tests in humans. Testing in mice can only pin down a likely starting point for experiments to determine a human dose, but the way in which you calculate that starting point is fairly well established for most cases. That established algorithm is essentially the same for most ingested and intravenously (or intraperitoneally in small animals) injected medicines, but doesn't necessarily apply to other injection routes — the relationship between different forms of injection, dosage, and effects is a complicated and surprisingly poorly mapped topic, and we'll set that to one side here. Some compounds, as always, are exceptions to the rule, and the only way that scientists discover that any specific compound is an exception is through testing at various doses in various species.

Given that, the discussion here should be taken to apply only to orally administered drugs, as that is the deliberately restricted scope of this outline. Further, when considering pharmaceutical dosage, it is important to emphasize that more is not better; this cannot be approached in the way people tend to naively approach the (over)use of dietary supplements. The primary goal, if self-experimenting, is to take as little as necessary of any chemotherapeutic, senolytic compound, as they are all toxic in any meaningful dose. I encourage a careful reading of the papers in which the side-effects in patients at chemotherapeutic doses and treatment durations are described, as well as the studies showing aggressive chemotherapy to produce a higher, rather than lower, load of senescent cells. That the dose makes the poison is an ancient adage, but no less true today.

The steps to figure out a suitable starting point for a human test of an orally administered senolytic pharmaceutical are as follows: firstly read the mouse studies for the senolytic compound in question, to find out how much was given to the mice and for how long. Doses for most ingested pharmaceuticals of interest will usually be expressed in mg/kg. Secondly, apply a standard multiplier to scale this up to human doses, which you can find in the open access paper "A simple practice guide for dose conversion between animals and human." Do not just multiply by the weight of the human in kilograms — that is not how this works. The relative surface area of the two species is the more relevant scaling parameter; read the paper and its references to understand why. Again, note that the result is only a ballpark guess at a starting point in size of dose. The duration of treatment translates fairly directly, however: for the period of treatment, start with the same number of doses, spacing of doses, and duration as takes place in senolytic studies in mice.

If there isn't enough data to do more than guess at a dose, then that is a good indication to write off that particular compound. Wait for more data, or look for different compounds with better existing data.

In the case of the dasatinib and quercetin combination, the mouse study of senolytic effects used a single dose of 5 mg/kg dasatinib plus 50 mg/kg quercetin. For a 60 kg human this scales up to a little less than 25 mg dasatinib and 250 mg quercetin. For comparison, mouse studies of dasatinib as a chemotherapeutic can be found that use 50 mg/kg per day for multiple days to evaluate its ability to kill cancer cells. A very useful study on dose effects and duration for dasatinib in humans used a dose of 100 mg in volunteers, and you can find other trials of dasatinib as a cancer treatment at that dose. Quercetin is an established and widely sold supplement, and it would be a real challenge to consume enough of it to cause any ill effects, never mind significant ones, judging by the toxicology data.

Another way of thinking about dosage is to aim at producing the same concentration of the pharmaceutical in blood that was used in cell culture studies, or observed in mouse studies. To do this you will need existing human data on how a dose maps to concentration in blood or tissues. The senolytic mouse study noted above and the useful human study provide sufficient numbers to make an estimate:

in the cell culture section of the mouse study, 100–200 nM/L is the effective concentration — more than that adds no greater benefit. Given the molecular weight of dasatinib, one can convert the observed blood concentration of 104.5 ng/ml in the human study for a 100 mg dose to get something in the ballpark of 200 nM/L. The human study also provides results for a 180 mg dose if you want to try scaling up or down to estimate the dose needed to hit different concentrations.

So via one method, the single dose for a 60 kg human is 25 mg dasatinib and 250 mg quercetin. Via the other method the single dose is in the vicinity of 75–100 mg dasatinib and 750–1000 mg quercetin, assuming that we scale up the quercetin to match the dasatinib, and depending on where we are aiming for in the 100–200 nM/L concentration. The second dose is at chemotherapy trial levels, but is a single dose rather than taken daily over weeks or months, so the impact will accordingly be more limited. Remember that there is no evidence to suggest that dosing with a senolytic treatment like the dasatinib and quercetin combination frequently will achieve any better result than dosing once every few years: the treatment kills the senescent cells it is capable of killing, and until more of those cells are created in significant numbers, more of the treatment will most likely do nothing helpful. There are apparently senolytic self-experimenters out there taking dasatinib regularly; I think this shows a poor understanding of the situation, and is probably harmful.

Because there are no published senolytic studies using venetoclax, coming to some kind of ballpark human dose for that purpose involves analogy and educated guesswork. The approach is to compare cancer studies of navitoclax and venetoclax, of which there are many, and then scale the venetoclax cancer study dose down in accordance with the difference between the cancer and senolytic study doses of navitoclax.

Firstly, the senolytic dosage for navitoclax in mice from the 2015 study is 50 mg/kg daily for two periods of 7 days spaced 14 days apart

— one thing you'll notice fairly quickly in all of this data is that BCL-2 family inhibitors compare unfavorably to dasatinib in terms of the amount needed and duration of treatment. That translates to a 60 kg human dose of around 250 mg.

A good place to start researching comparative dosages for venetoclax and navitoclax is the 2015 summary paper "ABT-199 (venetoclax) and BCL-2 inhibitors in clinical development." From there, and the references, the chemotherapeutic human dose of navitoclax was settling to somewhere in the 200–300 mg per day range for 14 to 21 days before it was discarded in favor of other tools, with the upper end of that dose range producing the aforementioned ugly side-effects related to platelet loss. Dosage for human cancer trials of venetoclax is, on the other hand, all over the map: doses range from 200 mg to 1200 mg daily carried out over a period of a few weeks to a month, with the dose given cycling in sometimes complex ways.

So if one were forced to put a pin into the map based on these numbers, the senolytic human dosage of venetoclax would probably be in the 400–600 mg/day range, every day for a week. Note that this is firmly in chemotherapy-with-side-effects territory, and there is no direct supporting evidence for effectiveness in mammals whatsoever. All said and done, the rough back-of-the-envelope estimation comes to a result that looks very unattractive, given that we expect better options in the future.

Alvespimycin is an HSP90 inhibitor, and this class of pharmaceutical may produce senolytic effects through less direct BCL-2 family inhibition; the effects and side-effects look broadly similar to those of navitoclax and venetoclax. One thing to note when researching this compound is that it is used in both injected and oral forms, so one has to be careful to work with only the papers that cover the oral delivery mode. The senolytic study in mice used an oral dose of 10 mg/kg provided three times every other day, which scales up to something like 50 mg with the same dosage schedule for a 60 kg human.

For comparison, a mouse cancer study used dosages in the 5–15 mg/kg range, with a variety of daily and intermittent schedules. A human cancer study also used a wide range of doses, and is a good resource if you are interested in reading up on the potential side-effects; based on their data, the authors recommended either 40 mg every other day or 20 mg daily for a period of four weeks of every six weeks for the follow-on study. Once again, this edges into chemotherapy territory, though less severely than venetoclax; anyone considering it would be making a personal call with too little data behind it to talk sensibly about risk.

Near all studies of chemotherapeutics start with low doses, a tenth of the expected study dose, and near all of them report a couple of patients who experience enough of a reaction at those low doses to drop out or require adjustment of the protocol. If risking chemotherapeutics yourself for senolytic purposes, even single doses or doses for a short time only, it is still important to first test a low dose at a tenth of the desired level or so, to help ensure there is no adverse reaction. As is true for all of the rest of the considerations here, if you try this, it is entirely your own responsibility to identify, understand, and manage the risks involved.

Obtaining Senolytic Pharmaceuticals

The mechanics of sourcing compounds from overseas manufacturers are common to any such self-experiment: find small gold-badge suppliers on Alibaba, pay by wire transfer, and receive shipments at a business mailbox rather than a residential address, accepting that customs will occasionally confiscate a package. The point specific to senolytics is to buy roughly twice what you think you will need, split across at least two suppliers in widely separated regions, since some of each batch goes to validation. Quercetin needs none of this: any specialist vitamin store will sell it, or it can be ordered online.

Storage of Senolytics

Dasatinib, venetoclax, and similar compounds are manufactured as fairly resilient powders and then formed into pills where sold as medications. In powder or pill form, put them in airtight containers in a fridge, and they have a shelf-life of a few years; the specific storage recommendations are easy enough to find online. The same is true of quercetin. This is one of the big advantages of most ingested pharmaceuticals versus injected pharmaceuticals: they are comparatively low-maintenance, stable, and long-lasting, meaning less logistical planning and effort.

Validating the Purchased Senolytics

A compound that turns up at the door may not be what was ordered, or may be impure, so each batch is worth checking by liquid chromatography and mass spectrometry against the reference data for the pharmaceutical — or rather, worth paying a small lab company, found through a marketplace such as Science Exchange, a little over $100 a sample to do it. Handling powders adds one wrinkle: weighing out a 5 mg sample from each order calls for a jeweler's scale such as the Gemini-20, small spatulas, sealable vials, and labels. Label each sample, keep a record linking it to its supplier, and triple-wrap it for shipping.

Ingestion Logistics for Powders

To match the way in which ingested compounds are taken in most studies, in pill form, it is probably best to make up pill capsules rather than just, for example, taking a measure of the powder in water or wrapped in bread. This is fairly easy to manage, given the tools already obtained for measuring out small powder samples. Specialist vitamin stores, and a range of other vendors, sell empty gelatin pill capsules for supplement enthusiasts, and they will do just fine here. Putting powder into capsules is a fiddly business that only becomes more frustrating with age; I'd suggest trying it out with flour if you haven't done this before — it is a lot harder than one might think. Fortunately, there are a variety of simple, inexpensive tools to help with that. At the very least, unless you happen to have three hands, a capsule holding tray is essential, and I'd recommend some form of small powder funnel.

Establishing Tests and Measures

Unfortunately there is no established, proven, useful test that can directly assess senescent cell level in humans or human biopsies. It is possible to use immunohistochemistry to assess cellular senescence in tissue samples, a standard approach in animal studies, but no one appears to have yet validated that in humans, given biopsies taken from a living individual. Since senescent cells are generated temporarily by wounding, it is quite possible that anything that starts with a biopsy will prove to be unhelpful as a before-and-after comparison measure for senolytic trials — the levels measured may not bear any resemblance to the normal levels absent a wound.

Without a direct measure, we must fall back on indirect assessments of the detrimental effects of senescent cells. The objective here is a set of tests that anyone can run without the need to involve a physician, as that always adds significant time and expense. Since we are really only interested in the identification of large and reliable effects as the result of an intervention, we can plausibly expect a collection of cheaper and easier measures known to correlate with age to be useful. Once that hill has been climbed, then decide whether or not to go further — don't bite off more than is easy to chew for a first outing.

From an earlier exploration of likely tests, I picked the following items on the basis of a likely connection to the actions of senescent cells, reasonable cost and effort, and ability to carry out the test without a physician's office being involved. That leaves the tests below quite focused on the cardiovascular system, particularly measures influenced by vascular stiffness, and inflammatory and other markers in the bloodstream:

- A standard blood test, with inflammatory markers.
- Resting heart rate and blood pressure.
- Heart rate variability.
- Pulse wave velocity.
- Biological age assessment via DNA methylation patterns.

Those cardiovascular measures respond to exactly the sort of change in blood vessel elasticity a senolytic treatment would be expected to produce. Positive movement in these metrics is also achievable through months of physical training, so a short-term change following treatment would be a meaningfully interesting data point.

There exist online services such as WellnessFX where one can order up a blood test and then head off the next day to have it carried out by one of the widely available clinical service companies. Of the set of test packages offered by WellnessFX, the Baseline is probably all that is needed for present purposes, though it's worth shopping around.

A simple but reliable tool such as the Omron 10 is all you need to measure heart rate and blood pressure. Worth noting here are a couple of general principles for cardiovascular measures. Firstly, the further away from the center of the body that the measurement is taken, the less reliable it is — the more influenced by any number of circumstances, such as position, mood, stress, time of day, and so forth. Fingertip devices are convenient, but nowhere near as useful as something like the Omron 10 that uses pressure on the upper arm. Secondly, all of the above-mentioned line items also influence every cardiovascular measure, so when creating a baseline or measuring changes against it, carry out each measure in the same position, at the same time of day, and make multiple measurements over a week to gain a more accurate view of the state of your physiology.

Surprisingly few of the numerous consumer tools for measuring heart rate variability actually deliver the underlying values used in research papers rather than some form of aggregate rating derived by the vendor; the former is required for any serious testing, and the latter is useless. Caveat emptor, and read the reviews carefully. As an alternative to consumer products, some of the regulated medical devices are quite easy to manage, but good luck navigating the system to obtain one; the easiest way is to buy second-hand medical devices via one of the major marketplaces open to resellers, though that requires a fair-sized investment in time and effort. After some reading around the subject, I settled on the combination of the Polar H10 device coupled with the Selfloops HRV Android application, and also gave the EliteHRV application a try. Despite the many recommendations for Polar equipment, I could not convince either setup to produce sensible numbers: all I obtained during increasingly careful and controlled testing was a very noisy set of clearly unrealistic results, nowhere near the values reported in papers on the subject. Plenty of people in the quantified self community claim that these systems work reasonably well, so perhaps others will have better luck than I did. Take my experience as a caution, and compare data against that reported in the literature before investing a lot of time in measurement.

For pulse wave velocity, choice in consumer tools is considerably more limited. Again, carefully note whether or not a device and matching application will deliver the actual underlying data used in research papers rather than a made-up vendor aggregate rating. I was reduced to trying a fingertip device, the iHeart, picked as being more reliable and easier to use than the line of scales that measure pulse wave velocity. Numerous sources suggest that decently reliable pulse wave velocity data from non-invasive devices is only going to be obtained by measures at the aorta and other core locations, or when using more complicated regulated medical devices that use cuffs and sensors at several places on the body.

Still, less reliable data can be smoothed out to some degree by taking the average of measures over time, and being consistent about position, finger used for a fingertip device, time of day, and so forth. It is fairly easy to demonstrate the degree to which these items can vary the output — just use the fingertip device on different fingers in succession and observe the result. A good approach is to take two measures at one time, using the same finger of left and right hand, as a way to demonstrate consistency. While testing an iHeart device in this way, I did indeed manage to obtain consistent and sensible data, though there is a large variation from day to day even when striving to keep as many of the variables as consistent as possible, meaning that only sizable effects could be detected.

DNA methylation tests can be ordered from either Osiris Green or Epimorphy / Zymo Research, though the latter takes a fair few weeks for delivery. From talking to people at the two companies, the normal level of variability for repeat tests from the same sample is something like 1.7 years for the Zymo Research test and 4.8 years for the Osiris Green test. The level of day to day or intraday variation between different samples from the same individual remains more of a question mark, though I am told they are very consistent over measures separated by months. Nonetheless, as for the cardiovascular measures, it is wise to try to make everything as similar as possible when taking the test before and after a treatment: time of day, recency of eating or exercise, recent diet, and so forth.

An example of one approach to the daily cardiovascular measures is as follows, adding extra measures as a way to demonstrate the level of consistency in the tools:

- Put on the Polar H10; this is involved enough to increase heart rate a little for a short period of time, so get it out of the way first.
- Sit down in a comfortable position and relax for a few minutes.
- Measure blood pressure and pulse on the left arm using the Omron 10.

- Measure blood pressure and pulse on the right arm using the Omron 10.
- Measure pulse wave velocity on the left index fingertip over a 30 second period using the iHeart system.
- Measure pulse wave velocity on the right index fingertip over a 30 second period using the iHeart system.
- Measure heart rate variability for a ten minute period using the Polar H10 and Selfloops.

Over the course of an experiment, from first measurement to last, it is important to maintain a consistent weight, diet, and level of exercise. Sizable changes in lifestyle can produce results that may well prevent the detection of any outcome resulting from a first generation senolytic pharmaceutical using the simple tests outlined here. Further, when taking any measurement, be consistent in time of day, distance in time from last exercise or meal, and position of the body. Experimentation with measurement devices will quickly demonstrate just how great an impact these line items can have.

Guesstimated Costs

The costs given here are rounded up for the sake of convenience, and in some cases are blurred median values standing in for the range of observed prices in the wild.

- Business mailbox, such as from UPS: $250 / year
- Baseline tests from WellnessFX: $220 / test
- MyDNAage kits: $310 / kit
- Osiris Green sample kits: $70 / kit
- Omron 10 blood pressure monitor: $80
- Polar H10 heart monitor: $100
- iHeart monitor: $210
- American Weigh Gemini-20 microscale: $90
- Miscellaneous equipment — spatulas, labels, vials, pill capsules, etc.: $60

- 2 x 2g orders of dasatinib via Alibaba: $300
- 2 x 5g orders of venetoclax via Alibaba: $1300
- 2 x 2g orders of alvespimycin via Alibaba: $400
- Store-purchased quercetin capsules: $10
- Shipping and LC-MS analysis of samples: $120 / sample

Schedule for the Self-Experiment

The process of discovery, reading around the topic, ordering materials, and validating the pharmaceuticals should be expected to take a couple of months. Once all of the decisions are made and the materials are in hand, pick a start date. The schedule for the self-experiment is as follows:

- Day 1–10: once or twice a day, take measures for blood pressure, pulse wave velocity, and heart rate variability.
- Day 10: bloodwork and DNA methylation test.
- Day 11–15: test a 1/10 dose of the senolytic compounds used, one by one, and abandon the effort if issues are experienced.
- Day 16: start to carry out the program of treatment.
- Day 31–40: repeat the blood pressure, pulse wave velocity, and heart rate variability measures.
- Day 40: repeat the bloodwork and DNA methylation test.

The exact timing is not really important, but it is a good idea to allow enough time following the end of the dosage for things to settle down. In animal studies, senolytic effects occurred fairly rapidly, as did the benefits, but allowing a few weeks of time in a human self-experiment still sounds like a good idea. Certainly it costs nothing to take that step.

Where to Publish?

The case for publishing your results rather than keeping them to yourself, and for doing so safely even anonymously, is the same as for any self-experiment. For senolytics in particular, a positive change in a basically healthy person in their late 40s or early 50s may be too small to separate from measurement noise — but a negative result of that kind is still worth reporting, since it helps steer better future efforts.

Final Thoughts: Why Not Wait?

Given all of the cautions above, why not wait? Waiting can be a very sensible strategy. The state of senolytic therapies is progressing rapidly. New and less chemotherapeutic senolytics are emerging, such as FOXO4-DRI. At some point in the next few years, reliable direct tests for senescence will arrive on the scene, allowing a much better view of whether or not these treatments are actually achieving the claimed results. That said, it doesn't hurt to plan, and it doesn't hurt to tinker with some of the component parts of a plan. That is how we can determine whether or not it is worthwhile to experiment now versus waiting to experiment later with better tools.

The Underground

Fight Aging!, 27 July 2009

Looking back from the perspective of 2035, I guess we should all be surprised that it took so long. The Vegas Group came together formally sometime in 2016, though the first kick-off meeting was the year prior, at one of the bi-annual conventions for longevity research held in California. By that time, more than a dozen gene manipulations and other biotechnologies had been shown to significantly extend life in mice, but no progress was being made to develop these technologies for human use. The Vegas Group was a natural outgrowth of a decade of advocacy and anticipation for human enhancement technologies — technologies that no regulatory regime in the US or Europe was going to allow anyone to develop, still less deploy.

There were initial fractures in the Vegas Group around the course of political change versus direct action, but by 2017 the direct action contingent consisted of about a hundred people all told. Their declared objective was a distributed collaborative effort to develop human versions of the most successful longevity and metabolic enhancements demonstrated in mice, and to cultivate hospitable medical groups in the Asia-Pacific countries. When these technologies were developed, the founding members would cast lots and carefully test upon themselves, in rotation, through the agency of medical centers in Asia. In doing this the hope was to spur change in the public view and greater progress in the commercialization of these technologies — and of course to gain access to manipulations that were greatly extending life in mice. "Pulling the big red lever," as one of the founders said, a venture where altruism and greed collide to best effect.

By this time, biotechnologies had become cheap enough to enable a growing amateur development community, akin to the hackers of the 1970s and the open source movement of the 1990s and onward, and it is this community, stirred up and cultivated by the core Vegas Group, that ensured success. It was a challenge, a middle finger to US authorities who were at that time attempting to shut down the open biotech movements, and a tangible way to prove that the "priesthood of the universities is done and gone." Of course, members of that priesthood pitched in to help, some to the detriment of their careers, others clandestinely.

By 2019, the first round of therapies took place amongst the founding Vegas Group members. This happened to some local fanfare in Singapore, milked for effect by the sponsoring parties. About half of the modifications to genes and repair procedures for the damage of aging were successful, as shown by assays and testing, but no serious side-effects occurred — a lot of prestige in the broader amateur biotech community rested on getting that part right at least. So there, in 2019, you have some of the first humans walking around with replaced mitochondria, cleaned-out cells, manipulated myostatin genes, addition of bacterial genes to eliminate obesity, and so forth. Strangely, it met with less interest in the mainstream press than you might expect — the pop-sci and scientific press release services wouldn't touch it.

Really, that's when the accident should have happened, but with the benefit of hindsight I think that people became complacent. The seventeen deaths in 2021 were avoidable — some of the Vegas Group were emboldened by earlier successes and rushed a new discovery, underestimating the risk. By that time more than fifty people — a core of the original Vegas Group, new members, and new offshoots of the group elsewhere in the world — had been successfully modified in ways demonstrated to extend life or improve the functioning of biochemistry in mice, and were leading normal lives. A few of the older group members had died, but not of any cause linked to their experimental treatments. The methods for making these modifications to humans were freely available online, and within the reach of perhaps a few hundred skilled amateur biotechnologists. They had promptly been outlawed by most European and the US regulatory authorities.

The commercial outgrowth of the Vegas Group was by this time also underway. Two companies affiliated with the original founding members, and two more working independently on the science, were striding towards commercial offerings of the technologies. None were based in the US, of course, and all but one were structured to take advantage of the trends in medical tourism from America and Europe. Even at ruinous early-adopter rates, as a luxury good for the wealthy, two of these companies went on to make their founders fabulously wealthy in a very short time.

Sadly, it was to be another decade before we came to where we are today — the first signs that US regulators might finally cease their prohibition of longevity-inducing and other enhancement technologies now offered widely in Asia and safely used by ever-increasing numbers of people.

From the perspective of the mainstream US scientific community, we still don't know if these alterations work as advertised on human longevity, and it may be 2055 before researchers are willing to move beyond measurement of markers and biochemical changes to acknowledge actual benefits to life expectancy from the procedures pioneered by the Vegas Group. Meanwhile, people are improving themselves.

Strangely, the Vegas Group has faded from view, a historical curiosity of the online cyclopedias, now supplanted in the "official" histories by the companies that rose up from their efforts. As a sign of success, being forgotten in the wake of the vessel you help to launch has something going for it — burying the daredevils is the first action of a successful industry setting up for the long term.

Interlude: 2019

Fight Aging!, 31 December 2019

It is that time again, an arbitrary midwinter point in the annual pilgrimage around the sun at which we take a look back to summarize some of the high points of the past year. As has been the case for a few years now, progress towards the implementation of rejuvenation therapies is accelerating dramatically, ever faster with each passing year. While far from everyone is convinced that near term progress in addressing human aging is plausible, it is undeniable that we are far further ahead than even a few years ago. Even the public at large is beginning to catch on. While more foresightful individuals of past generations could do little more than predict a future of rejuvenation and extended healthy lives, we are in a position to make it happen.

The State of Funding

A great deal of venture funding is arriving or preparing to arrive to support biotech startups working on means to treat aging. This year saw the launch of the Longevity Vision Fund, among others; I can think of three groups presently working to launch new mid-sized longevity-focused venture funds in 2020, and this is as seen from my fairly sedate perch of observation, without any great attempt to reach out and ask folk for a census. This activity will have a beneficial influence on public and private funding available for fundamental science in this part of the field. It also influences non-profit advocacy, as new organizations such as the Academy for Health and Lifespan Research are created. New governmental initiatives are also emerging, such as the Healthy Longevity Global Grand Challenge, and not just in the US: the UK government is putting out position statements on healthy longevity. Regulators are being petitioned by the scientific community to approve treatment of aging as a recognized goal for medicine. We should expect this trajectory to continue, though beyond the clearance of senescent cells, where development of therapies is very much a going concern, few approaches to rejuvenation are very close to making the leap from laboratory to clinical development.

There is no shortage of new companies targeting aging, many of which are, unfortunately I think, focused on manipulation of stress responses rather than rejuvenation. BHB Therapeutics works on ketosis mechanisms. Turn.bio aims to produce a safe way to partially reprogram cells in vivo, restoring youthful function. Samsara Therapeutics works on autophagy enhancement. Rejuvenate Bio works on gene therapies to slow aging, initially in dogs. My own company, Repair Biotechnologies, still young, raised a seed round to fund our work on thymic regeneration and reversal of atherosclerosis. The Oisin Biotechnologies spinoff OncoSenX also raised a seed round this year to deploy their suicide gene therapy in cancer patients, and LIfT Biosciences raised funds to develop immune cell transplants that have been shown to do very well against cancer in animal models. Underdog Pharmaceuticals is the SENS Research Foundation spinout targeting 7-ketocholesterol, which may be important in more than just atherosclerosis; they raised a seed round this year and are well on their way. Nanotics has launched a senolytics program based on interfering in the mechanisms senescent cells use to evade immune surveillance. The recently created Aging Biotech Info service now maintains a curated list of companies in the longevity industry, for anyone wanting the fuller picture.

Speaking of funding, the SENS Research Foundation's year-end fundraiser brought in more than three quarters of a million dollars this past year — money that supports the most effective form of altruism available, given the scale of the potential benefit. The SENS Research Foundation remains one of the most important organizations in research aimed at treating aging. Just because senolytics to clear senescent cells are a going concern, we cannot ignore the fact that the rest of the rejuvenation research agenda is nowhere near as advanced. It still needs funding, and near all funding for many of these vital projects remains philanthropic. We fund it. We are the people who make that difference, ensuring that important research projects can advance to the point at which they attract the support of more conservative, mainstream sources of large-scale funding.

Conferences and Community

These days, I'm as often as not out and about in the world raising funding or reporting on progress for a startup biotech company, Repair Biotechnologies, and found at many more conferences than would otherwise be the case. Side-effects of the growth of the longevity industry over the past few years include a change in the tenor of existing scientific conferences, as well as the addition of new conference series on aging that are focused as much on industry as on academia. This past year, I attended and wrote up notes on the SENS Research Foundation / Juvenescence gathering in San Francisco, held alongside the big JPM Healthcare conference; the first Longevity Therapeutics event, also in San Francisco; the Longevity Leaders conference in London; the vitally important Undoing Aging in Berlin; Biotech Investing in Longevity in San Francisco; the Ending Age-Related Diseases conference organized by LEAF in New York; BASEL Life, Founders Forum, LSX USA, and Giant Health in quick succession later in the year; the Alcor New York Science Symposium;

and the Longevity Week events in London coordinated by Jim Mellon and his allies.

Many conference presentations and interviews with members of the growing community were published over the year, too many to note — a measure in itself of how far the field had come.

Clinical Development

Drug development pipelines are moving forward, though not always smoothly — there is a high failure rate in the development of medical biotechnology. Eidos Therapeutics announced phase 2 results for their approach to preventing transthyretin amyloid aggregation. GenSight is presently struggling with phase 3 for allotopic expression of mitochondrial genes: the mechanism works, the earlier trials passed, and now reaching sufficient efficacy is proving to be a challenge. Intervene Immune published interesting results from their small thymic regeneration trial, while Libella Gene Therapeutics is launching a patient-paid trial for telomerase gene therapy. The resTORbio approach to inhibition of mTORC1 failed a phase 3 trial for immunosenescence, which may or may not cast a pall over that part of the industry. The TAME clinical trial for metformin, using a new composite endpoint as a surrogate for aging, was funded this year and will start soon — this despite the point that metformin remains a terrible choice of intervention, picked because the FDA couldn't object to it on technical grounds, not because anyone thinks that it will produce meaningful results for patients. Opinions are mixed on this topic.

The first human trials of senolytic therapies to clear senescent cells reported results this year, starting with promising results for lower dose dasatinib and quercetin versus idiopathic pulmonary fibrosis. Data from an as yet incomplete trial of dasatinib and quercetin versus chronic kidney disease has confirmed that these senolytics do clear senescent cells in humans in the same way as in mice. Unity Biotechnology announced results from their first trial of senolytics for osteoarthritis of the knee, and is moving on to phase 2, though there are those who think there is still a long road ahead to the clinic. A trial of fisetin by the Mayo Clinic has yet to publish results, but the Forever Healthy Foundation published a risk/benefit analysis covering what is known of fisetin as a senolytic in the meantime.

Cellular Senescence

Senolytics remained the brightest spot: as a class of therapy they look about as close to a panacea as it is possible to be, and new evidence over the year pointed to potential treatment for a long list of conditions — Alzheimer's, osteoporosis and osteoarthritis, atherosclerosis, pulmonary and cardiac fibrosis, kidney disease, diabetes, and more. The gap that still needs closing is measurement: there is as yet no good way to quantify the burden of senescent cells in a living person without a biopsy, and until there is, it is hard to know exactly how much any given treatment achieves.

Everywhere Else

Beyond senescence, the year brought incremental progress on nearly every other front of the damage-repair program at once: clearing the malformed mitochondria that falter with age; the long-delayed effort to break the sugary cross-links that stiffen tissue; a growing recognition that chronic inflammation, not amyloid alone, drives neurodegeneration; in vivo cellular reprogramming; work on the aging gut microbiome; the rapid proliferation of epigenetic clocks as candidate biomarkers of aging; and a regenerative-medicine field moving from first-generation cell therapies toward extracellular vesicles and engineered tissue. Each is a chapter in its own right; together they are the sound of a field that has stopped arguing about whether aging can be treated and started working out how.

In Conclusion

A great deal of progress is being made in the matter of treating aging: in advocacy, in funding, in the research and development. It can never be enough, and it can never be fast enough, given the enormous cost in suffering and lost lives. The longevity industry is really only just getting started in the grand scheme of things: it looks vast to those of us who followed the slow, halting progress in aging research that was the state of things a decade or two ago. But it is still tiny compared to the rest of the medical industry, and it remains the case that there is a great deal of work yet to be done at all stages of the development process. Senolytics must reach the clinic and widespread availability, and that will involve the deployment of vast amounts of funding. At the same time, however, numerous other equally important lines of rejuvenation research are still largely stuck in the labs or in preclinical development at best. There is much left to accomplish.

PART VII

The Backup Plan

People who are preserved are not dead and gone, just dead, with a chance to return.

— Fight Aging!, November 2016

And if the therapies do not arrive in time? He has an answer, and he made it long ago: he signed up to be frozen. These chapters make the rational case for cryonics as a backup plan, and then, characteristically, turn the question on himself — whether he should stop writing and go work on cryonics instead.

Never Bet on One Horse

Fight Aging!, 12 October 2015

The primary plan for anyone in middle age or younger today remains what this book has already made the case for: push the science, outrun your own aging via actuarial escape velocity, and beat the curve before it beats you. You can't make yourself younger yet, but you can help to speed up the development process — it is certainly moving at far below optimal speed at the present time.

That is the primary plan, and for every primary plan there must be a backup plan. Never bet on just one horse. The backup plan for evading the end that comes with death by aging is cryonics: low-temperature preservation of the fine structure of the brain on clinical death. Cryopreservation organizations will maintain the data of your mind in its physical form for the decades it will take for restoration to active life to become a viable possibility. That will, at minimum, require near complete control over cellular biochemistry and regeneration, as well as a mature molecular nanotechnology industry capable of repairing broken cell structures, removing cryoprotectant from tissues, and similar tasks. None of these goals are impossible or unforeseen, it is just that the necessary technologies don't exist today. Preserved individuals have all the time in the world to wait, of course.

In order to be cryopreserved you have to undergo a very unpleasant set of experiences: you have to age and you have to die, and do so naturally with little help, since our backwards legal systems don't allow for assisted euthanasia in a way that can mesh with cryonics protocols and organizational procedures. Further, compared to remaining alive and healthy thanks to working rejuvenation treatments, cryonics will for a long time to come be a leap into the dark with an unknown chance at ultimate success. It is still infinitely better than any of the other choices open to the billions who will die too soon to benefit from near-future rejuvenation therapies.

Strangely, after four decades of organized operation cryonics remains a tiny, niche, non-profit industry, for reasons that remain unclear and much debated. Cryopreservation is certainly a far better option than the many strange things people choose to have happen to their bodies following clinical death, usually for no better reason than everyone else does it. Is it little more than the fact that you have to prepare some time in advance to make it cost-effective via life insurance? That reluctance likely shares roots with the broader cultural aversion to treating aging as a medical condition at all. At root, all it would really take for cryonics to become a dynamic and competitive industry is for more people to sign up and express interest.

In recent years the popular press have transitioned from ridicule to balanced respect on the topic of cryonics, and the level of attention has increased. I think at least some of this has to do with growing interest in treating aging as a medical condition, though the relationship may be indirect: people who influence opinions tend to support both life extension and cryonics research and development. In the past decade we've seen a growing acceptance of the transhumanist ideals for longevity and the defeat of death that were first discussed realistically and robustly over the course of the 1960s to the 1980s. Many more people are now on the inside of what was once a small intellectual circle, and visionary thinking from that time is now taken as a foregone conclusion for technological development. That said, journalists are ever journalists, and still largely miss the very important difference between freezing, which is something that cryopreservation seeks to avoid, and vitrification, which is the goal of modern procedures. Freezing produces ice crystals that are highly damaging to tissues, whereas vitrification minimizes that outcome.

The coverage that does exist tends to run the same course every time: interview the earnest applicants, tally up the bodies in their thermos-sized dewars, quote a fee schedule, and move on without ever engaging with the underlying science of whether preservation and eventual restoration might actually work. But arguing about optics is a poor substitute for making the rational choice. If your options are cryopreservation, with some unknown chance of eventual restoration, or guaranteed and permanent oblivion via burial or cremation, the case for signing up isn't complicated.

The Moral Urgency

Fight Aging!, 16 May 2011

Our descendants will look back on us with some horror: an era of peoples who could have largely saved themselves from oblivion, but didn't. Suicidal, negligent barbarians they'll call us — and they'll be right. We have had the tools to preserve the brains of the dead for the long haul for quite some time now, and preserve them well enough that the fine structures storing the data of the mind remain intact. Death is death, but oblivion only occurs when the present detailed arrangement of matter in your brain is destroyed. For so long as that data is preserved, there is the chance that future technology and circumstances will lead to a restoration of life, such as through the use of advanced nanotechnology and charitable groups dedicated to returning the preserved to active life once more.

The preserved have all the time in the world to wait, after all.

When we are stirred to think of it, we have a sort of horrified sympathy for our distant ancestors: peoples who lived brutish, painful, and shorter lives because they had no choice. Many thinkers of past ages envisaged a better world and better lives, with no hope of seeing that transpire with their own eyes. The same sentiment will be applied by our descendants to those who came a little way before us:

doomed generations who could grasp at ideals and possibilities, but had no practical way to achieve these ends.

Sympathy is appropriate for those with high ideals and no possibility of ever achieving them — simply born too soon. But what of those who could achieve but turn aside and shirk the task? That would be us. I'd say we and our immediate forebears became suicidal, negligent barbarians about ten years after it became possible to build an industry to preserve brains, with a cost at scale that was affordable to the masses — on a par with funerals, for example.

For preservation methods along the lines of plastination, that point probably came and went somewhere in the 1930s or early 1940s, with the growth of the chemical industry into a true giant. The 1930s also saw the development of the first electron microscopes, and life scientists of the time arguably possessed enough knowledge of the cell to make educated guesses about which forms of chemical or low-temperature preservation would retain the data of the mind.

Here we stand, three generations removed, a thousand distinct human cultures in which there is little to show that we desire anything other than oblivion or self-delusion when it comes to our lives. Judging by actions rather than words, people who greatly desire both lasting life and health are a minority indeed. And with each passing year, another fifty million lives vanish into the maw to be destroyed utterly.

Why Trusts Won't Save You

Fight Aging!, 18 June 2009

A recent meeting of members of the cryonics community turned up a useful discussion of one of the smaller, more esoteric fields of thought that relate to cryonics: how to preserve assets post-mortem, so that you won't be utterly without property when (if) you are restored by future medical technology. The discussion centered on asset-preservation trusts of the sort some cryonics organizations now arrange through specialist lawyers. The mechanics are arcane: only an irrevocable trust can be self-settled — funded by, and paying out to, the same person — and only a handful of US states permit those at all. The enthusiasts trade notes on which jurisdiction serves best, weighing South Dakota and Delaware for their long friendliness to trusts that never expire against their differences in state income tax and trustee fees.

I note that the cryonics community, rather like the diverse libertarian community, possesses a sizeable minority with a great (and I think misplaced) belief in the power of contracts — of words on paper. You see it in the constitutionalists in the US or the fellows looking for loopholes in tax laws that will enable them to escape the

IRS entirely. Words on paper, however, have only as much weight as there are economic incentives aligned with them.

When thinking about these matters, it's best to look at the incentives rather than the details. The trouble with property left undefended, as the ancient Egyptians and every other culture that buried wealth with the dead has handily demonstrated, is that no-one else's interests are aligned with yours. You'd like your wealth — in whatever form it happens to be in — to be your property once more when you are revived from cryonic suspension. Every other person in the world will benefit from taking these resources for themselves while you are out of the picture.

When you place your wealth with a trust holder, you have also given future representatives of that trust holder — and anyone else in a position to plunder or take advantage of the trust — an incentive to see that you are not revived. This is probably not all that important for a small trust held for only a few decades by a large and prideful organization, but the larger the collected mass of trusts and the more time that passes, the less likely it is that the original terms will be honored.

Worse, under the scenario whereby cryonics becomes a large industry — and the practice of establishing asset preservation trusts becomes equally large — those folk being stored are collectively creating organizations that have as much of a financial incentive to use the political process to block revival efforts as they do to play nice so as to obtain new customers. Not to mention being a fat target for others to plunder via that same political process.

(The cryonics providers themselves have the opposite incentives: they want to be seen to provide good service, so as to obtain new customers. They also want the customers who have paid a lump sum on suspension to be revived as soon as possible so as to free up resources for new customers. The former is much more important than the latter.)

My own assumptions on the likely time to revival — ignoring for a moment all the slings, arrows, and unknown odds between here and there — are based on revival only being likely when it is also cheap enough for small self-interested groups to undertake. If you look at historical development cycles for new technologies, and take note of where researchers presently are with biotechnology and nanotechnology, someone who is cryopreserved prior to 2020 might be waiting 60 years or more until the economic circumstances favor revival. That's a long time for any agreement to last absent the people who originally signed it, especially when money is sitting around untended.

Why I Signed Up

Fight Aging!, 16 July 2016

I've long considered the cryonics industry to be a sensible, necessary undertaking in an uncertain world — the fallback plan for those of us racing rejuvenation research towards actuarial escape velocity, in case we don't personally make it in time.

Uncertainty is one of the reasons why advocacy and fundraising is so important. It is also the reason why a backup plan is a good thing to have. The only viable backup plan for the foreseeable future is cryopreservation, which is to say the low-temperature storage of the brain following clinical death in the hope of future restoration. The evidence is good for vitrification of brain tissue to preserve the fine structure of tissue that encodes the mind, and while the process in practice can always be improved in quality and organization, the (currently unknown) odds of survival following cryopreservation are infinitely higher than those attending any other end of life choice. After the grave there is only oblivion, but for so long as the data of the mind is preserved, there is hope that sufficiently advanced medical nanotechnology will one day bring you back.

I have put off signing up for cryopreservation for a decade or so. This isn't uncommon; after all, it involves paperwork, adult responsibility, planning ahead, thinking about unpleasant events, and all that.

People put off many other things for these and similar reasons. Writing wills, buying houses, getting married, starting companies, and so on. That doleful feeling of some unknown scope of paperwork that will have to be accomplished in the event that you do get your act together and set forth to be a responsible adult is ever a strong deterrent. Still, sooner or later all these tasks have to be carried out, and while no-one enjoys wading through legal documents, it is never as bad as you think it is going to be. If you are unfamiliar with the process of signing up for cryonics organization membership and cryopreservation, let me tell you that it is much less work than buying a house. It is about two and a half times the work of getting a life insurance policy, if that helps calibrate things any better. Typically, it runs as follows:

- Contact one of the established cryonics providers (Alcor or the Cryonics Institute in the continental US) and ask to join. They will send you a raft of paperwork typical of membership organizations, but since it involves the highly regulated area of end of life decisions you will have to have some of the documents witnessed and notarized. You are signing up to pay a monthly fee to help keep the lights on, and to donate your remains to the provider on death, but this is all provisional pending organization of payment.
- Obtain life insurance to pay for your eventual cryopreservation. This is where you will want advice, as there are only a few life insurance companies that have experience with this sort of thing. I recommend chatting to Rudi Hoffman, who has acted as an insurance agent for scores of people in the cryonics community. Paying for cryopreservation with life insurance involves establishing a policy that will pay out to the cryonics organization on death rather than to the more usual parties, such as family members. The matter of how large a policy to purchase is an interesting question — but the short of it is that more is generally better, as the future is uncertain. Life

insurance companies will want to pick through your medical records, will send a nurse to check your vital signs, and will usually find some reason to charge a little more than the early, optimistic quote that was given sight unseen.

- Then you wait for paperwork to cross the country back and forth, as everyone involved between you, the cryonics provider, the insurance agent, and the insurer has to sign everything. It might take a month or two, depending on how many of the end points are comfortable with electronic document exchange, but there is usually a lot of back and forth required.
- With membership and contracts in hand, you should take care of a few other supporting legal documents. This means making a will that determinedly and in no uncertain terms states your wish to be cryopreserved, and just as importantly establishing what is known as a living will or advance healthcare directive, a declaration of who you want to have power of attorney in the case of your incapacity, and what actions they should take. That should obviously be someone you trust to ensure that you are cryopreserved, and who for preference has absolutely no financial interest in the outcome one way or another. Because the contents of the will and living will are somewhat non-standard, you will probably have to make use of an actual living, breathing attorney rather than one of the very helpful standardized form legal services that have sprung up of late.

There are cautionary tales in the cryonics community regarding people who trusted their immediate family to ensure that they were cryopreserved, and those family members then sabotaged the arrangements in order to access the life insurance funds. This has happened on numerous occasions. Much as I might sound a misanthrope to say this, the best assumption to keep in mind when setting up the legal aspects of your cryopreservation is this: when on your deathbed, everyone not employed by the cryopreservation organization is a potential adversary, interested in the funds that they might be able to obtain for themselves at the cost of your survival. It doesn't matter how well you know people today, or the nature of your relationships with them. Cryopreservation is (hopefully) decades in your future, half a lifetime or more away: people change, and people who don't even exist today will be involved in your end of life care, having legally meaningful ties to you by kinship. Fortunately, it is possible to arrange matters to make it very hard for even family members to sabotage your arrangements, and most of these approaches are well documented by the cryonics providers.

For example, firstly you should ensure that no-one in your family is even named in the life insurance policy you are buying. It should pay out to the cryonics organization, and you might pick an entirely unrelated charity to benefit should you be unfortunate enough to die in a way that prevents your cryopreservation — thus the incentives align for the cryonics provider to try their best, as they get nothing if they fail, and your family has no interest in the outcome either way. Secondly, persuade as many close family members as possible to sign and have witnessed affidavits of support for your cryopreservation. Third, structure your will and living will to ensure that there are material incentives for your family in the event that you are successfully cryopreserved, and there are no material incentives for any of them to block that outcome. Fourth, record your determination to be cryopreserved, both in your will and in other media, and keep that reasonably up to date. Lastly for this list there is the matter of avoiding the unwanted attention of the state in the form of autopsy or other delays after death. The best methods here vary by jurisdiction, but in some states registering a religious objection can work well.

The best way to keep everyone on your side in the matter of cryonics is to ensure that all of the incentives align with the outcome of a successful cryopreservation, should it come to that. The moment that someone can benefit by making it hard to achieve that goal —

that is when the problems start. Relying on the essential goodness of human nature has never been terribly effective when it comes to expecting people to follow through with something that they themselves do not support, or think is frivolous, or which requires them to sacrifice their own gain. If you are in the room and exerting your will, that is one story, but when you are gone, it is quite another.

AN OBITUARY

Fight Aging!, 5 December 2014

I rarely write obituaries, because once you start where do you stop? Perhaps a hundred and fifty thousand lives are lost every day, most due to aging and its consequences, and it isn't just the few people you happened to exchange emails with who are worthy of notice. Yet monuments are at root a selfish undertaking on behalf of the living, and we can easily bury ourselves in mourning and symbolism. Ultimately one has to ask: is this an initiative about death or is this an initiative about life? The world has too many thinly disguised death cults. Cruelly, even after yet another individual in one's personal circle of vision succumbs to the frailty of age all of our lives go on as before. We're still here with the same old to-do list in front of us — or at least we will be until we are not. But that is rather the point: we want to eliminate this part of the human condition, build the medical technologies to repair the breakages that cause aging and thus prevent all of its attendant suffering and death.

I've long admired the oldest people in this community. They participate with no hope at all of benefiting personally from the technologies they support: that is true altruism. It will be, I'd think, twenty years under even the best of circumstances before comparatively crude first generation rejuvenation treatments as envisaged in the SENS proposals become available. If you are in later life it is vanishingly unlikely that you will survive for long enough to benefit meaningfully from present research. Yet that research must happen. Someone must be first to benefit, and someone must be last to miss out.

So we get to this news from the Gerontology Research Group (GRG), providing notice of the death of their founder and organizer in chief Stephen Coles, a researcher and advocate for longevity science. This had been expected, I think, given the details of his ongoing public battle with cancer. He took full advantage of having a rough timeline at the end to ensure a good cryopreservation: he died of complications of pancreatic cancer at 73, having traveled to Scottsdale, where Alcor is based, to be close to the cryonics foundation in his final days.

There is an obituary in the LA Times. For as long as I've been involved in advocacy, Coles has networked with fellow researchers and gathered data on late age survival: he tracked the world's oldest people for over twenty years, and performed autopsies on more supercentenarians — people aged 110 or older — than any other pathologist, identifying TTR amyloidosis as a predominant cause of death among them. With his connections as a hub the GRG mailing list became a cosmopolitan watering hole at which gerontologists, other researchers, and advocates with many varied views on aging and medicine debated points and rubbed shoulders. In recent years Coles' own work helped to shape the SENS Research Foundation strategy of funding potential treatments for TTR amyloidosis. This condition may be a true final limiting factor on human life span, killing those who survive everything else. Or at least it will be until therapies exist, and the development of those therapies is presently underway — though, as always in matters related to aging, with too little funding for truly rapid progress.

Cryonics is the sensible choice for anyone finding themselves in Coles' position. It is the only presently available shot at making death a hiatus rather than oblivion, and it is one slice of the grand self-destructive tragedy of the modern human condition that next to nobody chooses this path. Restoration isn't impossible, just very hard. But instead all those lives, all those individuals, are lost.

To change this state of affairs many more respected people at the hub of their own networks of influence must make a very public choice to be cryopreserved. If we want to see a world without frailty and disease in aging then more people have to speak out and act accordingly. Each of us can only be the one person in this parade, of course, but congratulations and thanks should pass to Coles for choosing to be that one on both fronts.

And that must stand in place for the numerous obituaries and mentions I could have written in recent years. As the community grows there are ever more older members and so more people vanishing over time from the mailing lists and blogs. But what can you do? Fifty years ago you could do nothing but wish. Today, however, you can make a material difference: support the research, support the organizations, help to speed our progress towards the day on which people stop suffering and dying of old age.

Cryonics or Rejuvenation?

Fight Aging!, 3 March 2022

The small, long-standing cryonics community and industry is focused on saving lives by offering the possibility of low-temperature storage at death, using cryoprotectants to induce a state of vitrification rather than straight freezing, a shot at preserving the structure and data of the mind for a future society capable of revival from this state.

I was recently in New York to attend the 50th anniversary gathering for the Alcor Life Extension Foundation, among the oldest of cryonics organizations. It was an occasion also marking the launch of a new book by Robert Freitas, a specialist in molecular nanotechnology and nanomedicine, whose new work catalogues in detail what would be required to revive a vitrified individual.

People interested in progress in cryonics tend to be interested in progress towards the development of rejuvenation therapies, though there is too little overlap in the other direction. Both were once more fringe than they are now, but work on rejuvenation has expanded and found far more support than is the case for cryonics. There is an old debate, often held between those with a foot in both camps: given that most of us in the later half of life expect the likely pace of progress in rejuvenation therapies to result in a sizeable improvement in old age, but not prevent us from dying from the effects of aging, why are we not fully focused instead on enabling cryonics to become a robust, large, dynamic industry? Isn't the primary object to avoid death, to avoid ceasing to exist? Why not then work on the better solution to that specific goal, rather than a path that can only improve health and thereby longevity in the time we have left?

I was asked this again at the Alcor event by Emil Kendziorra, co-founder of Tomorrow Biostasis, a comparatively new European cryonics venture that may well be pushing some of the older organizations to modernize more than they would otherwise have done, in his role as provocateur. He makes a habit of this; it is a part of his advocacy for the cause.

While I can't talk to the decisions of others, I can shed some light on why I am presently working on rejuvenation, and only plan to focus more of my time on cryonics later in life. Pre-pandemic, I said — to Emil Kendziorra, even — that ten years from now I would be spending more time on cryonics. It may still be ten years at this point post-pandemic, if early access to reprogramming or a combination of other plausible technologies pans out over the next five years and proves influential on health. My timing to switch to work on cryonics is in one sense driven by my projected health trajectory. At what point in the future do I predict that I only have 20 years left before there is a real risk of serious, debilitating age-related disease that drastically limits my ability to contribute? At that point I should put a great deal more effort into making those 20 years a productive time for the cryonics industry.

More generally, I will put more work into cryonics when the balance tips in its favor. "Balance" is a loose word: guesstimation and gut feel goes into (a) what I think can be achieved for many people over the next few decades via improvements in rejuvenation therapies, (b) what I think can be achieved for myself 20–30 years from now in rejuvenation by helping now, (c) what I can do for the cryonics industry now versus later, and (d) the likelihood that my present efforts in the biotech sector will produce significant personal capital to devote to cryonics research, development, and industry growth.

That last point is an important one. Improvement in the practical implementation of low-temperature cryopreservation proceeds at a very slow pace, with minimal funding. This is a consequence of it being a small field. Yet we can argue quite strongly that the lack of demonstrated capabilities — such as, let us say, reversible vitrification of organs used in the tissue engineering and organ donation fields — is the biggest impediment to convincing the world that cryonics is real.

At this point, it would probably cost $10 million to $20 million to push reversible vitrification of organs past the point at which more funding will arise organically and industry will inevitably form. Either philanthropy focused on academic-style programs or a deep-pocketed venture backed company might achieve the same result. But money doesn't grow on trees, and so far even the more visionary philanthropists have committed only a fraction of this amount to this sort of cryonics research. Turning up to put my shoulder to the wheel with the intent to find this funding is one thing. Turning up with those funds in hand is quite another. The odds of either path working out, and when they will work out, are worth considering when running the calculus of when to become more involved.

In any case, I am sympathetic to the argument that one should be working to speed up the growth and development of a future cryonics industry. One day I will do more than talking about it. Just not today.

PART VIII

What It Is All For

How many summers are there in a life? How many victories? Countless when they lie ahead. All too few when they lie behind.

— Fight Aging!, May 2018

Strip away the science, the money, and the strategy, and a simpler question remains: why does any of this matter? Here is the personal and philosophical core of the book — the emergency rooms, the deaths, the arguments about identity and meaning, and his own plain answer to why he has spent his life on this.

Waking Up

Fight Aging!, 17 May 2008

One day, you wake up to realize that a particularly vital assumption about the world is wrong. Everyone who buys into it is wrong. Which is almost everyone in the world. Everything in the world that depends on it is wrong. Which is almost everything in the world. Now what?

The reconfiguration of realization is not unlike passing through a slow instance of the stage of booting up in the morning. (Who am I? What am I doing today? How did I get here? What are my axioms? Oh, right, got it.) The "oh, right" part might take longer, however, as pulling out one of the pillars of your constructed world generally means radical change. We humans are hardwired to hate change almost as much as we hate failing to share the same worldview as our peers.

But there's no going back from realization. Either you proceed according to your new understanding of the world, and figure out just what that implies, or you become good at living a large and uncomfortable lie. Embrace revelation or run from it.

So, one day, you wake up to realize that aging is the worst bane suffered by humanity, and the people of the world sleepwalk through an unending holocaust of suffering and death caused by the decay of their bodies. Furthermore, it quickly becomes apparent that large-scale work over a few decades will plausibly lead to medical technologies that prevent age-related frailty, disease and death. Acceptance of aging in this circumstance is like a slow-motion mass suicide, day after day after day.

But still it goes on, the grinding of the cogs, the great wrong that everyone just goes along with as though it were nothing.

Now what? You've seen behind the curtain. You have a secret that no-one you know takes seriously. Yet, you realize with clarity that nothing else you ever do in your life will be as important to the future — your future, humanity's future — as any time and effort you put towards bringing forward the defeat of aging. That's going to be hard to put back into the box, trust me.

The Emergency Room

Fight Aging!, 21 November 2014

Yesterday, I had occasion to spend six hours or so in the emergency room of a medical center largely focused on treating serious conditions that are most prevalent in old people. A part of that experience by necessity involved listening to the comings, goings, and conversations of those present. These are not private places: they are typically divided visually by screens but with no way to avoid overhearing the staff and patients.

So, by proxy, one gets to experience small and somewhat wrenching slices of other people's lives. It is very easy for even those who follow aging research and speak up for rejuvenation treatments to forget just how hard it is to be very old. It's one thing to know about the catalog of pain, suffering, and loss of capabilities, the conditions we'd like to find ways to turn back, and another to watch it in action. It is, really, a terrible thing to be frail.

A fellow was brought in a little while after I arrived, a 90-something man who looked a lot better on the exterior than perhaps your mental picture of a 90-something individual might be. Tall, and surprisingly lacking in wrinkles, stretched out on the rolling gurney under blankets, a mess of cables, and an oxygen mask. That he had fallen was what I heard from the conversation of the medics, and he was in pain. He cried out several times as he was moved from the gurney. It took some time and care to do it without hurting him more, given his weakness.

He seemed confused at first, but that was just my misperception: you try being 90 and in pain some time and see how well you do while you're being moved around and told to hold this and let go of that. The fellow answered the bevy of questions the receiving staff had for him, but the thing that caught at me was the time he took with the answers, and the questions he just missed. He was coherent, even quite sharp at times, not on any more painkillers than a handful of Tylenol, as I later heard, but he clearly struggled with something that we younger folk all take for granted: parse the question, find the information, form up a reply and speak it. Cognitive ability in all these areas becomes ever less efficient with old age, and there's something hollowing about hearing what is clearly a capable guy set back for a dozen seconds by a short question about one of the details of his fall. The medic repeated the question a few times and in different ways, which was clearly just making the information overload worse.

It sticks with you to be the observer in this situation and clearly and suddenly realize that one day that faltering older person will be you, trying and often failing to force your mind into the necessary connections rapidly enough for the younger people around you. I know this, but knowing it and having it reinforced by being there are two very different things. An aged person is no less intelligent, far more experienced, wiser and all the rest, but the damage to the structure of the brain that occurs even in those without dementia means that making use of all of that in the way it deserves is near insurmountable.

The fellow's 60-something daughter arrived a little later to provide support and fill in more of the details. A story was conveyed in bits and pieces: that he was near blind now, and just about too frail to walk safely, even with a frame. The blindness explained a great deal of what had sounded to my ignorant ears as confusion in the earlier part of the fellow's arrival. The fellow lived with his wife still, and she was of a similar age to him. His wife was not there because she herself was too frail to be undertaking even a short trip at such short notice. That seemed to me a harsh blow on top of the rest of what old age does to you. At some point you simply cannot do everything you'd want to as a partner. You are on the sidelines, and at the point at which your other half is most likely to die, you are most likely unable to be there.

In this case the fellow was in no immediate danger by the sound of it. By good luck this was in no way likely to be a fatal accident, but rather another painful indignity to be endured as a part of the downward spiral of health and ability at the end of life. Once you get to the point at which simply moving from room to room bears a high risk of accident, and this is by no means unusual for a mentally capable person in their 90s, then it really is just a matter of time before you cannot live for yourself with only minimal assistance.

When talking with his daughter while he waited on a doctor and medical assistants to come and go with tests and updates, the fellow was much faster in his responses, though this was interrupted by a series of well-meaning but futile attempts to ease his pain by changing his position, each as much an ordeal as the move from the gurney had been. The conversation between father and daughter had the sense of signposts on well-worn paths, short exchanges that recapitulated the high points of many discussions that had come before. She wanted her father to move into an assisted living facility, and this fall was the latest in a line of examples as to why it was past the time for this — she simply could not provide all of the support needed on her own. She wasn't even strong enough herself to be able to safely get him back up on his feet after a fall. He was concerned about cost and the difficulties of moving, uncertainties and change. They went back and forth on this for a while. "We have to accept that it's just going to be more expensive as we get older," she said at one point, and he replied "I think you're getting the picture now," and laughed. There wasn't much to laugh about, but we can all do it here and there under these circumstances. I believe it helps.

I walked out of there after my six hours of hurry up and wait was done. They were still there, and whenever it is he leaves to go home it is unlikely it will be on his own two feet. But this is a scene I'll no doubt be revisiting at some point in the future, some decades from now, playing the other role in this small slice of life. What comes around goes around, but I'd like it to be different for me, and more importantly to be different for millions of others a lot sooner than my old age arrives.

Which leads to this: why does Fight Aging! exist? Why do we do this? Why advocate, why raise funds for research programs into ways to treat aging that may take decades to pay off? We do this because we can help to create a future in which there will be no more emergency rooms like the one I visited, no conversations about increasing disability, no pain, and no struggles to answer questions as quickly as one used to. No profound frailty. All these things will be removed by the advent of therapies that can effectively repair the causes of aging, curing and preventing frailty and age-related disease, and the sooner this happens the more people will be spared.

Another Emergency Room

Fight Aging!, 8 August 2022

Recently, I had the occasion to make one of my very infrequent trips to the emergency room. As always the case to date, I get to walk out afterwards, after a very long period of hurry up and wait. Not everyone is so fortunate. One of the things one tends to find in emergency rooms is old people. So many more of life's slings and arrows become an emergency when one is frail, and old people are increasingly frail. Fall over? Emergency room. Sudden infection? Emergency room. And so on and so forth.

Nurses and doctors are inordinately overworked. Emergency rooms are a great place to watch the consequences of that in action. A hospital as an entity is caring in the aggregate. There are formal systems of triage, but a great deal more informal triage based on which of the human wheels are presently squeaking. People fall through the cracks in ways large and small.

Waiting is what one does, largely, in an emergency room. A great deal of waiting. Particularly if one walks in and has every prospect of walking back out again. The older woman across from me in the waiting room did not walk in. She was in a wheelchair, and frail to the point at which walking was out of the question. She was alone. The nurses had wheeled her out at some point after intake, and there she was, waiting like the rest of us. In her case, increasingly unhappy in the Stoic, quiet way of the elderly. The nurse had left her bag slung over the back of her wheelchair, in such a way as to be inaccessible to a frail older person, unable to apply the modest amount of strength to turn and lift it over. Trivial for you and me, impossible for her.

It was hard to tell that she was unhappy. It didn't show in her face. But after a few times of noticing that she tried to tug at the bag strap, and with no relative or friend in evidence, left alone, I went over to offer assistance. Perhaps others there might have had I not, but none did. I lifted off the bag, put it carefully in her lap, and left her to it. She rooted around, took out slippers and dropped them to the floor — which may as well have been on the other side of the ocean for her, inaccessible, and beyond reach. Then found her phone and started working with it. At least a frail person has that!

Unfortunately that turned out not to be the case. A little while later she caught my attention and asked me to call her house. She was difficult to understand, in part because of accent and the COVID-19 rules that lead to everyone still being masked in hospitals, but any conversation was difficult for her. She did not say much, and was slow with what she did say. It wasn't always clear that she understood me. Still, she gave me a number, and I called it. It was out of service, I told her as much, and she seemed to grasp why it wasn't working for me. She then fumbled with her memory, half-trying variations on the number, but not completing any of them.

I asked about her phone, a modern iPhone. Did she have the number in her address book? The phone had a lock code, the usual panel of numbers to enter. She tried that, as she had been, and the phone promptly locked her out for five minutes. Modern security at work. As we waited for that timer to complete, I talked to her, retrieved her slippers and put them on, as she indicated that this was desired. She did not really respond meaningfully to much else of what

I said. At one point, she told me clearly that she did not feel well. I flagged a passing staffer and asked him to find someone, and nothing came of that by the time the phone was accessible again. Caring in the aggregate!

I watched her try to enter the phone code again, and she did it in a way that strongly suggested that she did not recall the code at all, or was perhaps not grasping the nature of the lock screen, entering the numbers in ascending order until the iPhone locked her out again, for longer this time. At that point, I went to find an actual nurse myself, and wouldn't take no for an answer. To her credit, the nurse put away what she was doing and came out to see what could or should be done, and had the good idea to look in the intake records for a phone number to call.

The old woman was wheeled away, and I didn't see her again. I walked out somewhat later, the more fortunate and less age-damaged of the two of us. I am not a physician and cannot diagnose dementia, but aspects of the interaction were those of someone who no longer has the full function of their brain. Just considering the physical, she was frail to the point of being unable to support herself, but that in combination with mental deterioration, leading to no longer being able to recall a phone number or even work a modern phone, is a sobering thing to see. Left on her own, she was helpless, and someone had simply left her there.

This, fundamentally, is why we advocate for the treatment of aging. No-one should find themselves in the position of the old woman I met in that emergency room, a prisoner of her own old age, a shadow of who she once was, left alone and at the whim of those who cared only when prompted to do so.

On a Death

Fight Aging!, 1 June 2011

I had an artistic vision, a long time back, shortly after I had my initial realization with regard to engineered longevity, of the path to radical life extension as a sort of vertical ascension. I recall describing the vision of flight through darkness years ago, to my partner at the time — the rest of that conversation lost, along with much else, to failing memory. A room is packed with thousands of people, waiting. The walls and ceiling then unfold to reveal a void beyond, and the assembled masses begin to ascend towards some distant goal — arms spread and looking up in anticipation, as they fly upward, empowered by their own sheer force of will. The energies of their passage glow blue in the darkness, a thousand, thousand trails of light. But with each moment, some falter, and fall away into the abyss, their light-trails vanishing into the surrounding darkness — and the crowd thins and thins the further they ascend.

I am not an artist, needless to say, and so that vision will never make it much beyond the confines of my head. But it is an ugly reality of the human condition that only a fraction of us will have the opportunity to live through into the age of rejuvenation biotechnology — even under the most optimistic timelines, billions will die. Intellectually we can grasp this scale of death and destruction, but the ape within filters it out. He only cares about the people we associate with — and there aren't all that many of them, even allowing for the artificial and strange relationships of celebrity or much-loved fiction. The rest of the world? Just numbers.

That memory of people in flight through darkness is with me again because I learned today that my former partner died a few weeks ago, unexpectedly. As though the expectedness or the delay in learning of it should change anything of the meaning, but the ape within feels — strongly — that the details of tragedy are important. If we felt even a minuscule fraction of what we do under such circumstances for the vast crowds who fall from the flight each and every day, aged to death ... well, we'd be better than human, and aging would be much further along the path to a solution.

More than a hundred people have died since I started writing this, most of them from aging or the complications of aging. All just as much individuals as my former partner, just as much people with attachments, potential, things to do, desires and regrets. So very many regrets. But I feel only intellectual loss for that hundred, and not much of it at that — this is what it is to be built on top of a foundation of ape prehistory, to be able to put the terrible fates of countless others easily from our minds, but to be deeply troubled by the loss of one.

To make the best of being human, we have to do the important work, even though it will not reward us, even though the ape inside doesn't really give a damn. Being human is about not letting the ape drive, even though it's always far easier just to let him take the wheel. So no obituary and no talking to the dead to say things that perhaps should have been said one day — and now can't be. Hard as it may be. The only person who would be listening here is the ape inside, and talking to him is a pointless exercise.

The world, meanwhile, is lived in by the living — who are falling from the flight in great numbers, day after day, while too little is being done about it. When whatever happens to me eventually happens to me, I would hope that someone takes a few minutes to write exactly the same thing, and mean it.

A Copy of You Is Not You

Fight Aging!, 4 September 2012

I'm a first things first sort of a person. A tremendous amount of work remains on any path that leads to the creation of rejuvenation biotechnology capable of reversing aging — especially if we want it to arrive before we die of old age — so expending a lot of effort on thinking about what happens afterwards doesn't strike me as helpful. That said, it can't hurt to glance ahead here and there in order to anticipate the next array of possible challenges and endeavors.

So this is one of those short glances, focused on the narrow issue of swimming upstream against the culture of our time — which is a great deal more work and a great deal slower than going with the flow.

As with the broader resistance to longevity science described elsewhere in this book, human culture is reflexively conservative, slow to embrace anything new even when it is beneficial.

One day, the change that was hard-won will be the new normal and completely accepted, as it will be for longevity science and the defeat of aging. Until then, supporting rationality and faster development in engineered human longevity will continue to be harder than it should be.

This additional cost in time and resources imposed by the nature of our present culture is an existential threat. It threatens to kill us by ensuring that the development of effective ways to reverse aging in the old arrives too late. If we want to live, it's a fight we have to win — or at least convince a few tens of millions to become supporters of longevity science in the same way that most people are supporters of cancer research.

But let us look to the future, at what I see as a loosely analogous cultural battle that will start to arrive at around the same time as the means to reverse aging — one that will also present an existential threat to personal survival.

Consider that at some point in the next few decades it will become possible to simulate and then emulate a human brain. That will enable related technological achievements such as reverse engineering of memory, a wide range of brain-machine interfaces, and strong artificial intelligence. It will be possible to copy and alter an individual's mind: we are at root just data and operations on that data. It will be possible for a mind to run on computing hardware rather than in our present biology, for minds to be copied from a biological brain, and for arbitrary alterations of memory to be made near-immediately. This opens up all of the possibilities that have occupied science fiction writers for the past couple of decades: forking individuals, merging in memories from other forks, making backups, extending a human mind through commodity processing modules that provide skills or personality shards, and so on and so forth.

There is already a population of folk who would cheerfully take on any or all of these options. I believe that this population will only grow: the economic advantages for someone who can edit, backup, and fork their own mind are enormous — let alone the ability to consistently take advantage of a marketplace of commodity products such as skills, personalities, or other fragments of the mind.

But you'll notice I used what I regard as a malformed phrase there: "someone who can edit, backup, and fork their own mind." There are several sorts of people in the world; the first sort adhere to some form of pattern theory of identity, defining the self as a pattern, wherever that pattern may exist. Thus for these folk it makes sense to say that "my backup is me," or "my fork is me." The second sort, and I am in this camp, associate identity with the continuity of a slowly changing arrangement of mass and energy: I am this lump of flesh here, the one slowly shedding and rebuilding its cells and cellular components as it progresses. If you copy my mind and run it in software, that copy is not me. So in my view you cannot assign a single identity to forks and backups: every copy is an individual, large changes to the mind are equivalent to death, and it makes no sense to say something like "someone who can edit, backup, and fork their own mind."

A copy of you is not you, but there is worse to consider: if the hardware that supports a running brain simulation is anything like present day computers, that copy isn't even particularly continuous. It is more like an ongoing set of individuals, each instantiated for a few milliseconds or less and then destroyed, to be replaced by yet another copy. If self is data associated with particular processing structures, such as an arrangement of neurons and their connections, then by comparison a simulation is absolutely different: inside a modern computer or virtual machine that same data would be destroyed, changed, and copied at arbitrary times between physical structures — it is the illusion of a continuous entity, not the reality.

That should inspire a certain sense of horror among folk in the continuity of identity camp, not just because it is an ugly thing to think about, but because it will almost certainly happen to many, many, many people before this century ends — and it will largely be by their own choice, or worse, inflicted upon them by the choice of the original from whom the copy was made.

This is not even to think about the smaller third group of people who are fine with large, arbitrary changes to their state of mind: rewriting memories, changing the processing algorithms of the self, and so on. At the logical end of that road lie hives of software derived from human minds in which identity has given way to ever-changing assemblies of modules for specific tasks, things that transiently appear to be people but which are a different sort of entity altogether — one that has nothing we'd recognize as continuity of identity. Yet it would probably be very efficient and economically competitive.

The existential threat here is that the economically better path to artificial minds, the one that involves lots of copying and next to no concern for continuity of identity, will be the one that dominates research and development. If successful and embedded in the cultural mainstream, it may squeeze out other roads that would lead to more robust agelessness for us biological humans — or more expensive and less efficient ways to build artificial brains that do have a continuity of structure and identity, such as a collection of artificial neurons that perform the same functions as natural ones.

This would be a terrible, terrible tragedy: a culture whose tides are in favor of virtual, copied, altered, backed up and restored minds is to my eyes little different from the present culture that accepts and encourages death by aging. In both cases, personal survival requires research and development that goes against the mainstream, and thus proceeds more slowly.

Sadly, given the inclinations of today's futurists — and, more importantly, the economic incentives involved — I see this future as far more likely than the alternatives. Given a way to copy, backup, and alter their own minds, people will use it and justify its use to themselves by adopting philosophies that state they are not in fact killing themselves over and again. I'd argue that they should be free to do so if they choose. Nonetheless, I suspect that this form of future culture may pose a sizable set of hurdles for those folk who emerge fresh from the decades in which the first early victories over degenerative aging take place.

Immortality, Seriously

Fight Aging!, 23 January 2017

I'm not going to try to convince you that the foreseeable future is a wondrous place: either you accept the implications of the present rate of technological progress towards everything allowed by the laws of physics, in which case you've probably thought this all through at some point, or you don't. Life, space travel, artificial intelligence, the building blocks of matter: we'll have made large inroads into bending these all to our will within another half century. Many of us will live to see it even without the benefits of medical technologies yet to come: growing up without the internet in a 1960s or 1970s urban area will be the new 1900s farmboy youth come 2040. Just like the oldest old today, we will be immigrants from a strange and primitive near-past erased by progress, time travelers in our own lifetimes.

A century is an exceptional life for a human, but far greater spans of years will be made possible by the technologies of the 21st century. I'll plant a flag way out there on the field and claim a million years: a life of a length hard to envisage. I am an advocate for engineered human longevity, and I started on the path that led to Fight Aging! and related projects from the position that (a) immortality would be an unalloyed good if achieved, and (b) our understanding of cosmology does not yet rule out a damn good attempt at actual immortality — the "no death, ever" dictionary definition — or at least a life span of millions of years on the way to that end goal. If a million years is not long enough to figure out the aspects of the problem that cannot be answered today, I'm not sure what would be.

Living for a million years is a goal that can be envisaged in some detail today: the steps from here to there laid out, the necessary research and development plans outlined, and the whole considered within the framework of what is permissible under the laws of physics, and what the research community believes can be achieved within the next 20, 50, or 100 years.

Biotechnology is the first necessary step on this road of a million years: it is the gateway to the future. The future is only golden for you and me personally if we live to see it, and for many of us that will require rejuvenation biotechnologies like those worked on by the SENS Research Foundation. This golden future is one in which our biochemistry does our bidding, aging can be repaired, and molecular manufacturing is in full swing. It will be an age of bioartificial bodies, minds transferred to new and more robust mechanisms, artificial general intelligences, an end to most scarcities, and indeed, anything you might imagine that the laws of physics permit and enough time has passed to develop.

A philosophy of first things first is a good way to temper visions of the far future — and explains why I spend my time talking about rejuvenation biotechnologies, cryonics, and even basic common sense health practices that might stop you cutting a mere decade from your life expectancy. If I'd been born twenty years earlier, I'd have ended up primarily a cryonics advocate and volunteer. As it is, it looks like these first decades of the 21st century are the era in which the first rung on the ladder of simply remaining alive forever — which is to say building the means to continuously repair the biological damage of aging in these bodies of ours — can actually be achieved. If we can live another 50 years, grabbing a year here with good health and a year there through incremental advances in geriatric medicine, and if we can build a large enough research community interested in serious work on rejuvenation along the way, then we may live in restored youth and vigor for centuries longer.

If you project forward into the future based on today's accident rates, you'll find that an ageless human sustained by biotechnologies of cellular and biochemical repair has a life expectancy in the range of 1,000 to 5,000 years. Sooner or later that piano is going to fall upon your head hard enough that even advanced medical technology cannot fix your injuries in time. So the million year life span: how could that be achieved? The short and not terribly informative answer is that it will be accomplished by using advancing technology to dramatically reduce your vulnerability to fatal accidents, murder, and other unfortunate events that produce the same outcome. Once you start looking at living for even 100,000 years in much the same shape as you are today, it becomes apparent that almost any activity bears an unavoidable minimum level of risk that will jump up and kill you. Eating, swimming, walking ... breathing. Stretch out the timeframe far enough and the improbable and fatal will eventually occur.

The way past these risks is to change your form: your risk of fatality for any given activity is a function of your human physiology. Once the research and development community has achieved the goal of practical biotechnologies for the repair and reversal of aging, that will give us all a few hundred years of life in comparative statistical safety. Technological progress will continue across that long period of time, and I can't imagine that much of the toolkit needed for the next step in long-term risk reduction will remain beyond the capabilities of the human civilizations of the 2200s. Your own personal preferences for that next step will no doubt vary, but I would get my neurons replaced — slowly, one at a time over time, to ensure continuity of the self — with some form of much more robust, easily maintained nanoscale machinery. That allows for a range of new engineering possibilities: swapping out the body for whatever machinery of transport and support best minimizes risk; moving most of the business of life into a virtual world; physically separating my neurons while still remaining alive, conscious, and active.

It shouldn't be terribly controversial at this point to talk about machines that can do the job of a neuron, store all of the same information as a neuron, and integrate fully with surrounding real neurons. Researchers in recent years have assembled lobster neuron simulators from Radio Shack components, and made great inroads into manipulating the internal machinery of cells. These are toys and clunky barnstorming exercises in comparison to what lies ahead, but my point is that this is an active line of research, worked on by thousands of scientists and developers. Similarly, I would hope that interacting via virtual worlds and splitting up one's machine neurons between various locations follows fairly straightforwardly from having machine neurons in the first place. If your brain is made up of artificial neurons, why not throw in an internet connection, adjunct computer hardware, and encrypted wireless communication protocols?

Physical distribution of the self across many disparate locations is in fact the key point when it comes to considering risk over the long term. Locations have much the same issues with time, probability, and bad events as people do. Meteorites are a risk to consider, as are landslides, earthquakes, war, and volcanoes. The way to reduce your location-based risk dramatically is to spread out. You might imagine a wireless brain, using whatever the most robust communications technology of the time happens to be, scattered in a thousand separate machine bodies or vehicles across a continent, or even the whole planet. That might be good for many millennia of falling pianos of various types. However, once you start digging back into the geological and astrophysical history of the solar system, it becomes clear that spreading out over an entire planet still leaves you at risk on longer timescales. Probably not from impact events: I'll be surprised if humanity and its machine descendants fail to solve that problem within the next few centuries. But there will always be war, nearby supernovae, large solar flares, unusually massive volcanic events, and other unpleasant line items, however. Supernovae are the biggest of the known concerns, given that I expect it to be a long, long time before preventing them is a practical and ongoing business for the civilizations that follow man.

What to do about all of this astrophysical and grand geological risk? Spreading out is an option once again. Increase the size of your vehicles and neuron-machines to shrug off the worst case radiation projections for a nearby supernova. Provide them with the means to move about the solar system, and become a spacefaring entity, spread out over a sizeable selection of orbits. By that point in time, your physical presence resembles a small country of machinery, automation, and layers of delegation: perhaps you are a million heavily shielded self-powered containers and transmission systems distributed beyond Pluto's orbit. There is a trade-off for spreading out so far, however, and that is that you must slow down. The speed of thought is determined by the speed of communication between the neurons and sections of your brain. If your brain is light hours wide, you will live very slowly indeed — but with a life expectancy so long that you come out far ahead in the end.

There are other paths forward with varying degrees of risk. You might decide not to spread out, but rather live very fast by running your machine neuron brain on more capable hardware, for example; if you can pass a hundred years of subjective time in a year of real time then you have reduced your subjective risk for many fatal occurrences a hundred-fold. That would be a pleasant enough life as a part of a community of people all running at the same speed, and there is even room for technological development and research to occur at a fair pace under such a scenario. At present our still young computing technology is very, very far removed from the known theoretical limits on computational efficiency. There is a great deal of headroom for the approach of living more rapidly.

But to return to the immortality question: is immortality impractical? Given existing mortality rates and the uncertainties in the timeline for completing efforts to repair and reverse the damage of aging, it may be unlikely for many of us alive today. If progress is too slow, or we are simply unlucky in matters of health, then we won't get past the first step on the path. In other words, we will die — or at best undergo cryosuspension and its attendant risks — before the advent of sufficiently good rejuvenation biotechnology. As for the bigger picture, it is far too early to say whether immortality, the "no death, ever" version, is actually impossible. That requires further research into cosmology — so you might give it a million years or so and ask me again. Regardless, the slope of technology and possibility is curving up ahead of us to great heights, and it'll be a wild ride either way. Missing out on any of it would be a real downer, so why not spend more of your time and resources helping to get the first step accomplished? We should all support the development of rejuvenation biotechnology, as it is the gateway to a life that may ultimately prove to have few limits.

The Ethics of Paradise Engineering

Fight Aging!, 10 July 2018

The two strongest urges are firstly to seek pleasure, in all its myriad forms, and secondly to evade suffering, in all its myriad forms. The primordial glass half full and glass half empty of the human condition. These are the two sides of the hedonistic imperative, and are perhaps the most important motivations guiding the development of technology. Technology, and I use the word in its broadest sense, can satisfy these urges either by helping to eliminate suffering or by helping to induce pleasure. Technology to reduce suffering has throughout history largely consisted of the vast and complex fields of medicine and agriculture. On the other side of the fence, for the induction of pleasure, we find intoxicants and pharmaceuticals of other classes, as well as, arguably, every technological development that can be turned to conquest and control. Not all pleasures are good in the moral sense, or perhaps it is better to say that given our deeper origins in an animal world that runs red in tooth and claw, many of the chemical incentives inherent to our biology are triggered only through selfish and damaging acts.

There are nonetheless many pleasures that can be attained without causing harm or resorting to advanced forms of technology. Completing challenging work, triggering the evolved response to pattern and surprise that is humor, simply being present in an attractive location, participating in the puzzle palace of human interactions, physical or otherwise, and so forth. Altering the operation of our brains to induce pleasure without the need to undertake much of that work was a fairly early innovation, however. The point of much of technological progress is to achieve better results with less effort, after all. The logical end of that line is wireheading or a life science equivalent yet to be designed: an augmentation in the brain, a button that you push, and the system causes you to feel pleasure whenever you want. There are numerous other alternatives in the same technological genre that seem plausible, such as always-on happiness, regardless of circumstances.

This sort of thing makes many people nervous, and, sadly, rarely for useful reasons. One doesn't have to look much further than the continued efforts to make mood-altering drugs illegal to see a panoply of bad motivations and perverse incentives exhibited front and center. Not every drug user becomes an addict, and self-destruction through addiction is clearly something that people do to themselves, only aided by the drug. A drug is an enabling technology, like a hammer, and neutral in and of itself. There remains considerable uncertainty today over who is more or less prone to addiction, and why, though there are plenty of addictive games against which you can test yourself in that way, such as those in which the makers gleefully exploit the effects of variable reinforcement on the human mind.

That said, I suspect that even the most self-controlled of individuals has sufficient self-doubt to be wary of the advent of implementations of wireheading that might be, say, a hundred times better, cheaper, and safer than today's most influential mood-altering drugs. What would you do in the presence of that potential option to substitute for the hedonic treadmill of work and reward? Little Heroes by Norman Spinrad is a worthy, albeit partial, fictional exploration of that question, and I recommend it. It is worth asking yourself "why not gain control of my mind in this way?" — and then follow your own answers to their logical ends. That exercise will probably reveal a great deal about how you view the world and your place in it.

Reliable, safe, on-demand pleasure (or confidence, or feelings of well-being, or happiness) achieved through technologies such as wireheading is a topic that has been comprehensively explored in fiction and philosophy, and is very slowly trickling into the real world via some forms of early, only weakly effective pharmaceutical products. There is little I can add to that library here that hasn't been said well many times over. What I will say is that to my eyes this is actually the less interesting and less consequential of the two sides of the hedonistic imperative. It is the elimination of suffering, not the gaining of pleasure, that, when taken to its conclusions, will lead to a world and a humanity changed so radically as to be near unrecognizable.

Paradise engineering is a catch-all term for the creation and large-scale use of technology to build a world that satisfies the hedonistic imperative. For me at least, pleasure on demand is a nebulous, potentially dangerous, and much less important goal when compared to the concrete list of forms of suffering that we might address, especially given that for each of these pains and lacks we can envisage the necessary technologies and changes in some detail. The hedonic treadmill may well be inextricably tied to freedom in its purest sense: the freedom to make your own choices implies the freedom to make mistakes, and pleasure and its absence are an important part of the mental machinery by which we measure our use of time and effort. Suffering, however, is not necessary in this model. Thus I see the primary task of paradise engineers as being to take the list of suffering in some approximate order from greatest and most widespread harm to least and most localized harm, and work on solutions until such time as there is no more meaningful suffering. By this I mean solutions to the actual problems, the root cause of suffering, rather than any form of shortcut akin to wireheading or other forms of augmentation of feeling. Selectively blocking all unwanted physical pain is one response to the suffering caused by an incurable fatal degenerative disease, for example, but it isn't a good response.

It might surprise some that the greatest cause of human suffering is not the inhumanity with which all too many people treat one another, individually and collectively. Nor is it the related deficits in the organization of our societies: war, kleptocracy, repression, the enforced poverty that results when the bottom rungs of the ladder of growth are removed. The greatest cause of suffering receives the least attention. It is aging, the simple biological wear and tear of the body and the structures of the brain that support the mind. It affects everyone, and it causes drawn out pain, fear, and misery, alongside the loss of dignity, opportunity, and vigor — and ultimately the loss of the self as the mind decays. A staggering number of people are presently suffering in many ways because of aging. To the degree that we think of death as a loss and a form of suffering, then we should be prompted by the fact that aging is by far the greatest cause of death in our species.

This then, is the next goal for paradise engineering: to bring aging under medical control, and to reduce the cost of that control to the point at which everyone can live indefinitely in youthful health. It could be feasible within decades, given great enough support for the necessary research and development, as we all age for the same reasons, and identical mass therapies for billions could be produced with the greatest of economies of scale. Control of aging is not the first goal undertaken by paradise engineers, however, which is to say that paradise engineering has been taking place for some time. Many important incremental goals were achieved over the course of the past half century, for example: progress in agricultural technology sufficient to make famine impossible, save through human neglect and corruption; greater control over infectious disease; and many others.

After hunger and aging, there is still the other half of infectious disease to deal with, however. There are also the thousands of forms of internal failure of human biochemistry and biology unrelated to aging or infection. Ultimately the medical community seeks complete control over our molecular biochemistry, sufficient to eliminate all defects. At present a look to the future suggests that this goal will compete with the development of machine alternatives to biological systems, and humans will become hybrids of engineered, cultivated biology and artificial nanoscale machinery that will work with, enhance, or replace portions of our biology. Aging and disease will be banished, while malfunctions and breakages will be both far less common and cause only inconvenience when they do happen. We will have successfully defeated all of the most common sources of physical pain and dysfunction, either by remodeling the chassis of our biology, or by adding guards to protect it from harm.

Suffering is not only human, however. The natural world from which we evolved continues to be as bloody, terrible, and rife with disease as it ever was. Higher animal species are certainly just as capable of experiencing anguish and pain as are we humans, and the same is true far further down into the lower orders of life than we'd like to think is the case. We ourselves are responsible for inflicting great suffering upon animals as we harvest them for protein — an industry that is now entirely unnecessary given the technologies that exist today. We do not need to farm animals to live: the engineering of agriculture has seen to that. The future of paradise engineering could, were we so minded, start very soon with an end to the farming and harvesting of animals. That would be followed by a growing control over all wild animal populations, starting with the lesser numbers of larger species, in order to provide them with the same absolute control of health and aging that will emerge in human medicine. Taken to its conclusions, this also means stepping in to remove the normal course of predator-prey relationships, as well as manage population size by controlling births in the absence of aging, disease, and predation.

Removing suffering from the animal world is a project of massive scope, as where is the line drawn? At what point is a lower species determined to be a form of biological machinery without the capacity to suffer? Ants, perhaps? Even with ants as a dividing line, consider the types of technology required, and the level of effort to distribute the net of medicine and control across every living thing in every ecosystem. Or consider for a moment the level of technological intervention required to ensure a sea full of fish that do not prey upon one another, and that are all individually maintained in good health indefinitely, able to have fulfilling lives insofar as it is possible for fish. Artificial general intelligences and robust molecular manufacturing technologies, creating self-replicating machinery to live alongside and inside every living individual in a vast network of oversight and enhancement, might be the least of what is required.

At some point, and especially in the control of predators, the animal world will become so very managed that we will in essence be curating a park, creating animals for the sake of creating animals, simply because they existed in the past — the conservative impulse in human nature that sees us trying to turn back any number of tides in the changing world. It seems clear that the terrible and largely hidden, ignored suffering of the animal world must be addressed, but why should we follow this path of maintaining what is? What good comes from creating limited beings for our own amusement, when that same impulse could go towards creating intelligences with a capacity equal or greater than our own? Creating animals, lesser and limited entities that will be entirely dependent on us, to be used as little more than scenery, seems a form of evil in a world in which better choices are possible.

Given this, my suspicion is that when it comes to the animal kingdom, the distant future of paradise engineering will have much in common with the goals of past religious movements and today's environmentalist nihilists, those who preach ethical extinction as the best way to end suffering. Animals will slowly vanish, their patterns recorded, but no longer used. If animals are needed as a part of the world in order to make the human descendants of the era feel better, then that need can be filled through simulations, unfeeling machinery that plays the role well enough for our needs. The resources presently used by that part of a living biosphere will instead be directed to other projects.

That is the far future. Even now, however, we have the ability, the technological capacity, to eliminate more of the world's suffering than has already been tackled. All it would require is for people to make better choices. This hasn't happened because we are disorganized, because we are inhuman to one another and to animals, because evolved human nature produces harmful collective behaviors — the aforementioned war, kleptocracy, and so forth. Utopians of every stripe have had their say on how to fix these issues within the bounds of the human condition, but it seems quite clear that it cannot be done. The human mind, as presently constituted, results in behaviors and societies that will consistently sabotage the outcome of a paradise free from suffering.

Today human nature cannot be changed, but in the future we will become able to change ourselves, as well as to manufacture whatever nature we desire in the intelligences we create. Utopianism might be rescued by technology, by the ability to edit all of the fundamental aspects of human nature that we all presently take for granted. From a purely technical perspective, it seems feasible for a future society to engineer away the aspects of human nature that produce our failures: the urge to dominance, the urge to violence, the urge to cause pain and suffering and loss to others, as well as jealousy, poor impulse control, and many others. By the time that it is commonplace for organic brains to be augmented and replaced by other forms of processing machinery, I would imagine that the production of altered forms of human nature and human intelligence will be a going concern, alongside modeling and predicting the behavior of societies consisting of such modified human minds.

Unfortunately, I am skeptical that this technological capability can be successfully applied to solve the problem of human nature as it presently exists, even while it seems, from a purely technical point of view, possible to achieve the goal. Resources will always be limited at some level, whether atoms, energy, or computational power, as societies grow to match the bounds of usage. War, bad governance, and the like are the result of a race to the bottom based on violence and control of scarce resources. For so long as any section of society retains human-like nature sufficient to follow these historical patterns, then the rest will be driven to change or extinction, no matter how enlightened they are. To bypass this by changing everyone, imposing specific mental models on every living entity, would require a level of control and dictatorship that is hard to imagine coming to pass universally across the entire human polity — and that is quite beside the point that such an end is, in and of itself, just as malign as the behaviors it seeks to cure.

So: we can improve the state of lives, we can build a truly better world, but it will most likely always be flawed, marred by our own actions, just as is our world today. That is no reason to hold back, however, as there is so very much that might be improved even given that perfection is unattainable. Pain and anguish need not be our everyday companions, and need not exist throughout the animal kingdom. Suffering can be addressed, and every step we take in that direction is well worth the required effort.

Why I Do This

Fight Aging!, 5 August 2019

Much of the proceedings at Undoing Aging in Berlin earlier this year was recorded. I briefly escaped from the conference for an ad hoc, unstructured discussion with Adam Ford of Science, Technology, and the Future, who was interviewing as many people as he could during the event. It wound up a monologue on topics that were at the top of my mind at the time, particularly the present state of funding, the transformation of our community from a primary focus on advocacy and academic research to one in which a great deal of important work is now carried out in startup companies, and the utilitarian ethics of treating aging as a medical condition.

I'm Reason. I've been around in this community for quite the long time, going on I guess twenty years now, rather shockingly. I seem to have become old in my own lifetime; I'm not as young as I look, unfortunately. I run Fight Aging!, the blog, which I've used as a platform for advocacy for some time, the aforementioned going on twenty years, though more like fifteen now for that site. In 2018, I cofounded Repair Biotechnologies with Bill Cherman to actually jump into the industry and do some things. Prior to that I was investing as an angel in a few biotech startups focused on aging, where I felt it was a better choice than giving to the SENS Research

Foundation or other groups to do the research, because companies have the chance to attract a great deal more funding more rapidly than non-profits do, unfortunately — that is just the way of the world.

That is very much the transition of our community right now. And as that happens, I think it becomes much more important to think about why the hell are we doing this thing? Sudden influxes of vast amounts of funding are consequential. There are several hundred million dollar funds right now, focused on longevity, and there will be more next year, because it is a land rush right now. If you lose sight of why you are doing this, and thus what is the most effective approach, then you wind up with a bunch of idiots doing stupid things that won't work, and the upshot of that is that funding will be wasted. It is to a certain degree unavoidable — look at the dot com era; every new industry has its peak of hype, a bunch of stupid things happen, a bunch of charlatans come in and take funding from investors who don't know any better. It will happen, but I think that those of us who are here now, and have been here in this community, have something of a duty to try to reduce the size of that problem, down to some nominal minimum, to the degree that that is possible to achieve.

So why do we do this? The fundamental philosophy of the problem is that death is bad. Suffering is bad. That death is bad is the more debatable of those two. It is quite possible to construct an ethical position in which we say it is fine to be dead, you didn't exist for quite a long time prior to existing, and you will not exist for quite a long time after you cease to exist. That is the way of the universe; the Stoics were good with this position. But I think it is very hard to argue that suffering is acceptable, at least above the sort of "maybe I should get out of bed and do something today" level of suffering needed to motivate the human animal to go and do something constructive. Anything much more than that level of angst I think should go away — and certainly that includes the level of pain, physical decrepitude, loss of function, and horrible things happening to the people around you that comes with aging. That should go away.

The world just hasn't quite got there yet in terms of thinking about this in the right way. People think about malaria in the right way. Malaria is something like one six hundredth of the cost to humanity of aging, depending on how you want to measure all the little fripperies around the edges. So if we really feel up in arms about malaria, willing to spend billions on getting rid of it, which some people clearly are, then we should be spending trillions on getting rid of aging. We should be, but we are not, and that is why the advocacy community came into being. We have this weird mismatch between our capabilities and our goals. The world is a crazy place, I think everyone acknowledges that; there are many, many insanities that the human condition contains, be that politics, be that the way that some people like peanut butter, pick your poison. The present relationship with aging is just another one of these insanities: the world is insane with respect to aging, because accepting aging is insane. Why would anybody accept that he or she is going to crumble and die, and not want to do something about it? Plain, basic utilitarianism says that we should do something about it if we can do something about it. And we can do something about it!

So is the population of the world asleep at the switch because next to no-one realizes that we can do something about it? That will change pretty soon. Senolytics will wake up everybody. What if you can take one pill that makes your arthritis go away? That is basically what senolytics will do, when they are truly effective. The ones we have right now, that are available right now, appear to be fairly good at getting rid of arthritis, based on the results of trials yet to be published. Once this realization happens, I think there will be an interesting phase change. People will start to somewhat wake up from this business of "oh well, aging is just a fact of life, wherein we're all going to die horribly, let's just get on and try to paper over that." So no, instead let us go full on utilitarian and try to do something about it. That is essentially the philosophy of action here. It is that aging is so terrible that there is really no amount of effort that humanity could spend on this problem that would be too great. Of course we're so far away from anything that even approaches a reasonable amount of effort, given the level of death and suffering caused by aging, that for the foreseeable future we can keep on advocating and building hundred million dollar funds. If investors build another hundred of those hundred million dollar funds, that would be a nice start, but by no means the right amount of funding in order to reasonably address the problem, given what it actually costs.

The cost is enormous, and, sadly, most discussions about aging, when they do get going, really skate over the utilitarianism of it in favor of "wow, we're spending a lot of money on entitlements, we need to do something about this." That latter expression seems to be what passes for ethical thought in policy circles these days. It is a matter of "well, we're spending a bunch of funds, we should find a way to stop doing that." Then of course, the nihilists seem to be mostly in charge now, because their idea of spending less is to not treat old people for their conditions, rather than building rejuvenation therapies that stop old people from getting those conditions. As I said, it is insanity. This really just needs to change. So this is why the advocacy, and now that we're at the point at which funding can be raised for startup companies working on rejuvenation biotechnologies, these startups are just another form of advocacy, really, if you look at the bigger picture. We're not building therapies because we can do something with our small slice of the pie of aging, we're building therapies because if we show people that we can do something with our small slice of the pie of aging, then soon enough there will be another hundred companies over the next decade, working on their small slices. People will see success and attempt to replicate it themselves.

There are a hundred, two hundred, three hundred programs out there languishing in the research community that could be turned into companies, turned into therapies, doing useful things in and around aging. The research community is just not good at raising funding. They are not good at translating their research to the clinic. They are poor at a lot of things other than just advancing the science. I think it falls to the rest of us — where "the rest of us" means anyone who might be an entrepreneur, or in the venture industry, or an advocate — to set forth and sift through these programs, the output of the scientific community, and say "look, we should do something with these things." If the research community isn't pushing a program forward, well, this is a time in which anyone can wrap a company around a project, say "I think we can do something with this," go to the venture industry and get a few million dollars in seed funding.

That will be the case for the next twenty years, on and off, as the market cycles up and down. So we should have a thousand startup companies working on a thousand projects related to aging. While there are only seven categories of fundamental causative damage, per SENS, some of those categories are very, very broad in terms of their little individual components. We have to fix all of lipofuscin, and we don't even have a good catalog of everything that is in lipofuscin, just the major pieces of it. We have to get rid of all the amyloids, and that is a good few dozen projects right there. Replacing aged stem cells: a different cell population, different recipes for therapy for every tissue. And so on and so forth all the way down the list.

Then after we've worked through the SENS list of causative damage as it exists today, there will be all the things about aging that are problematic but are hidden by the fact that people presently die before they become problematic. Such as nuclear pore proteins in long-lived neurons. Some of those molecules never change after they are initially created. It is the same molecule for your entire life, and if it gets damaged, well, that is kind of a problem. How do we build the nanotechnology to go fix our nuclear pore proteins? That is a problem that no-one should much care about today, because there are fifty other things that will kill you before that will become an issue. But it will become an issue, eventually. If we come to live to 150, I'm willing to believe that your nuclear pore proteins becoming corroded and corrupted and reacted with is actually a serious issue, at least in neurons.

We can in principle replace everything except the brain. So the worst case scenario for the ultimate future is that they open your skull, take out your brain, and put it into a new body. I'm sure it will actually be somewhat more sophisticated than that, but this is just a thought experiment — what is the most radical thing you can make work in terms of replacement, in principle? That is moving the brain to a new body. What will probably happen instead is that your new body will be rebuilt from your old body: regeneration and rejuvenation by delivery of cells and therapies and control over cell behavior. But the brain itself? A really challenging problem, because you have to fix it without breaking it. I think we are along the way towards understanding the mechanisms to target for the early, preventative reductions in inflammation, to avoid supporting cells in the brain going crazy, to get rid of the protein aggregates. To try to keep things the way they were during your 30s. But that is just a starting point. There is so much to do after that. It is a big project. When I say trillions in funding, I'm serious. This is a very big thing, this is reinventing architecture when you are a caveman, going all the way up to the Renaissance, and building huge palaces. That is the scope of the project for us. I don't think it will take as long as it took the cavemen, but I think that we're definitely in for the long haul. To the extent that we can incrementally build meaningful rejuvenation therapies along the way, then many of us will also be in for the long haul, and this turns into someone's life's work. That life might be rather long.

I don't know how long people will live. I am in my late 40s, and if you can just run the thought experiment of the biotechnologies that will be available to me in my 80s, I won't look anything like an 80 year old person today. I will have no chronic inflammation; no senescent cells; probably no cross-links in my body; my stem cells will have been replaced; my immune system gardened; and so on and so forth through a long list of treatments that are going to happen in the next few decades, and are very plausible right now. So you can add these things up and say, right, if an 80 year old has no inflammation, no senescent cells, no cross-links, no atherosclerosis, what does that do to health? Do you still look like an 80 year old? Can you go run a mile? No-one knows, and we get to find out by doing it. That is the great adventure.

The big problems in aging are all comparatively simple to solve, and it is all benchwork in the lab to get your programs going. You don't need the massive computational, big data, machine learning projects that are popular right now. The only place where present artificial intelligence might be useful is in improving the state of small molecule drug discovery, and my belief is that small molecule drug discovery will go away, largely, in favor of gene therapy. So maybe your AI is looking for genes or proteins that are of useful effect, but the present process of finding genes that have useful effects is not terrible. It is having good results. The upshot is, ok, where do you use AI in this process outside of small molecule development? And I don't see anything in which AI is absolutely necessary, useful in any way other than incrementally improving the infrastructure, reducing costs. Targeting senescent cells with senolytics, that is where small molecules might be useful, but the best projects there don't involve small molecules. Dealing with mitochondrial DNA damage? No, that is benchwork, and it seems unlikely that small molecules can do anything meaningful there — that is gene therapy territory. Stem cells? Again, it is just a matter of developing the methodologies that can lead to successful therapies, and deep down under that development, you find a role for AI in anything where there is a lot of data to be analyzed, but it is only incremental improvement in cost and efficacy.

Infrastructure makes the world turn, and incremental improvement is not to be sneezed at, but it is just a part of the technology background. You can't just jump up and say "we're going to do AI for longevity", no. You are going to do AI for biotechnology in general, and biotechnology is then applied to longevity. So AI will vanish into the tool space. It won't be a major category that is up there on its own in the fight against aging. Right now it is because it is novel and because investors throw funding at AI like there's no tomorrow, and entrepreneurs and scientists follow the funding. So you get In Silico Medicine, for example, doing small molecule discovery AI, which is what most other people are now following on to do nowadays, because that is where the funding is in the present phase. But I think this will just fade into the background, it will be another tool in the toolkit. It isn't exciting, it is not category changing. It is an incremental advance, using computers a little bit more to help you do things when there is a lot of data involved.

Let's talk about Effective Altruism. That community is doing smart things, in the sense that Big Philanthropy is thoroughly corrupt, and one should ask the question: if I want to do good in the world, versus conning myself into thinking that I did good in the world, what should I in fact do? You don't give to the Red Cross, because the Red Cross is a thoroughly corrupt organization. The same for most large entities in philanthropy; they have enormous overheads, and most donated funding doesn't go to the projects you would want it to go to. The Effective Altruism movement in that sense is great, and a very overdue examination and critique of philanthropy as a whole. Secondly, if the effective altruists can find convincing ways to convince high net worth individuals to actually give sensibly, this will be a good thing. I suspect that the reason many high net worth individuals don't give sensibly is because they have absolutely no idea how to do good, and it is a big project to figure out how to do good at scale. If Effective Altruism can lead to more people in the high net worth category putting their funds into projects that actually have a good chance of improving the human condition, then that is a public good.

The third strand is obviously that the Effective Altruism community includes people who are quite concerned about which projects to fund from a utilitarian point of view, and to the extent that anyone takes even a cursory look at aging, it is obviously the case that aging is far worse than anything else that happens anywhere. Pick your favorite cause in the third world, and I can tell you that those people are suffering more from aging than from the target of your favored cause. Even for war, even for famine, it is still the case that aging is much, much worse. This is a sad thing, because we could be dealing with all of these issues, but when it comes to prioritization, yes, if you want to solve famine because it is terrible and causes people to suffer, then you also should be willing to work on solving aging in that same population, because it causes a far worse outcome to far more people. So the Effective Altruism community should logically work its way to advocacy for the treatment of aging as a medical condition, because it is undeniably the case that it is the worst problem facing humanity, and it is the most cost effective point of intervention to reduce suffering and death in the world. Even when intervening in tiny ways, the outcome is an enormous return on philanthropic investment in the cause.

So I think that the effective altruists do good, and I think that there are not enough of them, and I think that they are not talking about aging to the degree that they should. But they are coming at it largely from an outsider perspective, and except for a few, they don't understand the science, they don't understand the degree to which rejuvenation is possible. Effective Altruism is a young movement, it has a way to go yet, but it has the potential to be very important. We shall see how it develops. In terms of our community engaging with effective altruists, it is all just advocacy at the end of the day. To the extent that the aging research and development community needs funding, then we set forth and engage the effective altruists to the extent that they have funds, or can influence sources of funding. If it is more effective to talk to effective altruists, then we aging advocates will do that, trust me.

That might be challenging as an argument right now, as right now it is clearly more effective to talk to the venture capitalists, because they are very incentivized to put funding into these projects. Very large amounts of funding, in fact, to the point at which it would be very hard to raise that level of funding through any sort of philanthropic program. Unless of course you are talking to high net worth individuals. But convincing high net worth individuals to go and put funds into work on treating aging is Effective Altruism, whether or not you are cloaking it in that name. Certainly, I and others are guilty of poking high net worth individuals to say "have you thought about this a little bit? Do you want to get old? You can do something about it. So go do something about it." But it is an incremental process. You can't just flip a switch and have all of the trooping masses of the Effective Altruism community go off and spread the desired message. So we shall see. It will go where it goes.

There is an enormous waste right now in development and deployment of ineffective ways to treat age-related disease, those that don't target the causes of aging. Further it will cost a great deal to develop functional rejuvenation therapies that do target causes of aging. But if you look at the enormous amounts that are spent on merely coping with the consequences of aging, then making it go away is highly efficient. But of course it is not just about funding and cost, it is about effective reduction of suffering. Funding spent on anti-aging research is an enormously cost-effective way to reduce suffering, providing it is spent on the right anti-aging research, rather than the programs that are not likely to produce more than a small effect. So mTOR inhibitors are great compared to previous technologies for dealing with age-related diseases, because they influence a lot of processes, but the effect size is really not large in the grand scheme of things. If you are going to put funding into developing mTOR inhibitors, then fine, that is happening, then you should spend that same level of funding on aspects of the SENS program that can actually repair underlying damage, rather than trying to tweak the body to be a little more resilient to damage. People taking mTOR inhibitors are still going to die on the same schedule as the rest of us. They aren't that much more resilient. People with senescent cells removed, on the other hand, well, who knows. We will see what that does to life span. I think that the pensions and insurance companies are going to be in for a rude awakening. Personally, I think that five years of additional life is not an unreasonable guess, and that will break a lot of insurance companies if they haven't prepared successfully.

Regarding what will convince the world that meaningful progress is happening and further meaningful progress is possible, I think that recent developments in the laboratory, particularly around senolytics, are convincing to scientists. That is helpful. But I don't think that it convinces the world at large in a useful way. Things have to leave the lab for that to happen. The thing about senolytics is that even those initial compounds available now seem to be quite good at making a sizable impact on quality of life in older people, and possibly for autoimmune diseases, and a bunch of other things. To the degree that we can say "guys, we're giving you a rejuvenation pill, your arthritis is probably going to go away" and then if even half of the patients lose their arthritis, or their symptoms are greatly minimized, and they lose their other inflammatory conditions, and we turn back early Alzheimer's disease — and all of these are plausible things that senolytics should accomplish, based on the mouse studies — then if that happens, then suddenly rejuvenation therapies are a real thing, and people can stop saying it is impossible to rejuvenate humans. Then we can go from there to explain that this is just one part of a larger program. We're just doing one tiny thing, and not even that well, and look how good it is.

Senolytics will be the point at which an awful lot of things change. The early stages are happening right now. The self-experimentation community is doing interesting things with senolytics. Once the first studies that actually have large effects are published, it will be hard for regulators to keep these early senolytic drugs out of peoples' hands. There are 60 million people in the US alone who would benefit from senolytics because they are old enough to have conditions that are inflammatory. This should happen. It will happen. And that would be the moment, I think. Senolytics, not anything else. Aging is a huge burden. Effectively treating aging will solve many problems. Old people are old people because they are aged. If you rejuvenate them, then they are no longer old. They will have a better time of it, and if you have an aging population of 80 year olds who are physiologically like 65 year olds, then you have an aging population of 65 year olds, effectively. After that it is very easy.

Escape Velocity

Fight Aging!, 18 November 2015

Let us define immortality as being a state of agelessness, which seems a common colloquial usage these days. More precisely this means that the risk of death due to intrinsic causes such as wear and tear damage of vital organs remains the same over time, perhaps due to advanced medical interventions. Falling pianos are still going to kill you, and a hypothetical biologically young immortal in a hypothetical environment maintaining today's first world extrinsic mortality rate would have a half-life of 500 years or so, meaning that at any age, there is a 50% chance of evading a life-ending event for another 500 years. There are no human immortals by this criterion of a static intrinsic mortality rate, it seems. Immortality in a state of advanced frailty and coupled with a 90% or higher yearly mortality rate isn't the sort of circumstance that most people would aspire to, of course. It barely improves on the actual circumstance that the oldest of people find themselves in, all too briefly.

However, let us think beyond the box. Consider the small horde of children that you'll find playing and running in any junior schoolyard here and now. By the time the survivors of their cohorts reach a century of age, the 2100s will have arrived. If the current very slow trend in increasing adult life expectancy continues, adding a year of remaining life expectancy at 60 for every passing decade, then something like 25% of these present children will live to see that centenary. But I don't for one moment believe that this trend will continue as it has in the past. Past increases in life expectancy were an incidental side-effect of general improvements in medicine across the board, coupled with increasing wealth and all the benefits that brings. Across all of that time, no-one was seriously trying to intervene in the aging process, to address the causes of aging, or to bring aging under medical control. Times are changing, and there are now many groups building some of the foundations needed to create exactly this outcome, groups such as the SENS Research Foundation. The trend in longevity in an age in which researchers are trying to treat the causes of aging will be very different from the trend in longevity in an age in which no such efforts are taking place.

You don't have to dig very far into the state of the science to see that the first rejuvenation treatments are very close, their advent limited only by funding. If funding were no issue for senescent cell clearance, for example, it would absolutely, definitively be in clinics a decade from now. Other necessary technologies are more distant, but not that much more distant — the 2030s will be an exciting time for the medical sciences. For the occupants of today's junior playground, it seems foolish to imagine that by age 60 they will not have access to rejuvenation treatments after the SENS model at various stages of maturity, many having been refined for more than 30 years, at the height of their technology cycle, and just giving way to whatever radical new improvement happens next.

The sweeping differences and expanded technological capabilities that exist between today, the 1960s, and the 1910s are enormous. So very much has been achieved, and that pace of progress is accelerating. Those junior playground athletes of today will live to see a world even more radically different and advanced than our present time is in comparison to the First World War era. These are the immortals among us. The majority of them will have the opportunity to attain actuarial escape velocity, to keep on using ever-improving versions of rejuvenation treatments until they are gaining life expectancy at a faster rate than they are aging. Their cellular damage, the wear and tear created by the normal operation of metabolism, will be repaired as fast as it is generated. It is the rest of us, those of us who are no longer spring chickens, who are faced with much more of a race to the goal. The degree to which we can successfully fund and advocate the necessary research is the determinant of whether we can scrape by into the age of rejuvenation treatments, or whether we will gain modest benefits but still age to death — because we were born too soon, and because the rest of the world didn't get its collective act together rapidly enough.

The Last Mortals

Fight Aging!, 22 May 2019

With the development of rejuvenation therapies underway, and accelerating, somewhere ahead lies a dividing line. Some people will be the last to age to death, too comprehensively damaged for the technologies of the time to recover. Everyone else will live indefinitely in youth and health, protected from aging by periodic repair of the underlying cell and tissue damage that causes dysfunction and disease. Where is that dividing line? No one can say with certainty. I look at the children of today, with long lives ahead of them, and find it hard to believe that in a hundred years the problem won't be solved well in time for them to live for as long as they choose. Equally, people in middle age today will certainly benefit greatly from the advent of first generation rejuvenation technologies, such as senolytics, each narrowly focused on one mechanism of aging. Yet I'm skeptical that matters will progress rapidly enough to rescue them. So somewhere between those two points are the people on the very edge; the last mortals.

In a sense this isn't terribly profound at all. It is the same story for every as yet uncontrolled medical condition, where the medical research community is working towards effective treatments that will arrive at some uncertain future date. There will be those who are the last to die, just as the therapies that save everyone else are rolling out. It is only the magnitude that is greater in the case of aging — a hundred or a thousand times greater. Does the fact that it affects everyone mean that there will be public disorder, disputes between the first immortals and the last mortals, where only private, personal existential crises exist today? I think claims of societal unrest as a result of the realization that your children will live indefinitely, while you yourself will not, are likely overwrought. That, at least, is the shape of it: a difference of years, perhaps decades, standing between an ordinary human life and something far longer, decided by nothing more meaningful than the accident of when you happened to be born.

A Letter From the Future

Fight Aging!, 29 May 2018

I will be 85 somewhere in the mid 2050s. It seems like a mirage, an impossible thing, but the future eventually arrives regardless of whatever you or I might think about it. We all have a vision of what it is to be 85 today, informed by our interactions with elder family members, if nothing else. People at that age are greatly impacted by aging. They falter, their minds are often slowed. They are physically weak, in need of aid. Perhaps that is why we find it hard to put ourselves into that position; it isn't a pleasant topic to think about. Four decades out into the future may as well be a science fiction novel, a far away land, a tale told to children, for all the influence it has on our present considerations. There is no weight to it.

When I am 85, there have been next to no senescent cells in my body for going on thirty years. I bear only a small fraction of the inflammatory burden of older people of past generations. I paid for the products of companies descended from Oisin Biotechnologies and Unity Biotechnology, every few years wiping away the accumulation of senescent cells, each new approach more effective than the last. Eventually, I took one of the permanent gene therapy options, and then there was little need for ongoing treatments. Artificial DNA machinery floats in every cell, a backup for the normal mechanisms of apoptosis, triggered by lingering senescence.

When I am 85, the senolytic DNA machinery is far from the only addition to my cells. I underwent a half dozen gene therapies over the years. I picked the most useful of the many more that were available, starting once the price fell into the affordable-but-painful range, after the initial frenzy of high-cost treatments subsided into business as usual. My cholesterol transport system is enhanced to attack athero-sclerotic lesions, my muscle maintenance and neurogenesis operate at levels far above what was once a normal range for my age, and my mitochondria are both enhanced in operation and well-protected against damage by additional copies of mitochondrial genes backed up elsewhere in the cell. Some of these additions were rendered moot by later advances in medicine, but they get the job done.

When I am 85, my thymus is as active as that of a 10-year-old child. Gene and cell therapies were applied over the past few decades, and as a result my immune system is well gardened, in good shape. A combination of replacement hematopoietic stem cells, applied once a decade, the enhanced thymus, and periodic targeted destruction of problem immune cells keeps at bay most of the age-related decline in immune function, most of the growth in inflammation. The down-side is that age-related autoimmunity has now become a whole lot more complex when it does occur, but even that can be dealt with by destroying and recreating the immune system. By the 2030s this was a day-long procedure with little accompanying risk, and the price fell thereafter.

When I am 85, atherosclerosis is curable, preventable, and reversible, and that has been the case for a few decades. There are five or six different viable approaches in the marketplace, all of which basically work. I used several of their predecessors back in the day, as well. Most people in the wealthier parts of the world have arteries nearly free from the buildup of fat and calcification. Cardiovascular disease with age now has a very different character, focused more on failure of tissue maintenance and muscle strength and the remaining small portions of hypertension that are still problematic for some individuals. But that too can be effectively postponed through a variety of regenerative therapies.

When I am 85, there is an insignificant level of cross-linking in most of my tissues, as was the case since my early 60s. My skin has the old-young look of someone who went a fair way down the path before being rescued. Not that I care much about that — I'm much more interested in the state of my blood vessels, the degree to which they are stiff and dysfunctional. That is why removal of cross-links is valuable. That is the reason to keep on taking the yearly treatments of cross-link breakers, or undergo one of the permanent gene therapies to have your cells produce protective enzymes as needed.

When I am 85, I have a three decade patchwork history of treatments to partially clear this form of amyloid or that component of lipofuscin. Modified enzymes are delivered here, a gene therapy applied there. I will not suffer Alzheimer's disease. I will not suffer any of the common forms of amyloidosis that degrade heart muscle performance or disrupt function in other organs. The potential for such conditions is controlled, shut down with the removal of the protein aggregates that cause cellular dysfunction. There is such a breadth of molecular waste, however: while the important ones are addressed, plenty more remain. This is one of the continuing serious impacts to the health of older individuals, and a highly active area of research and development.

When I am 85, I am the experienced veteran of several potentially serious incidences of cancer, all of which were identified early and eradicated by a targeted therapy that produced minimal side-effects. The therapies evolve rapidly over the years: a bewildering range of hyper-efficient immunotherapies, as well as treatments that sabotage telomere lengthening or other commonalities shared by all cancer cells. They were outpatient procedures, simple and quick, with a few follow-up visits, so routine that they obscured the point that I would be dead several times over without them. The individual rejuvenation technologies I availed myself of over the years were narrowly focused, not perfect, and not available as early as I would have liked. Cancer is an inevitable side-effect of decades of a mix of greater tissue maintenance and unrepaired damage.

Do we know today what the state of health of a well-kept 85-year-old will be in the 2050s? No. It is next to impossible to say how the differences noted above will perform in the real world. They are all on the near horizon, however. The major causes of age-related death today will be largely controlled and cured in the 2050s, at least for those in wealthier regions. If you are in your 40s today, and fortunate enough to live in one of those wealthier regions, then it is a given that Alzheimer's, atherosclerosis, and the other great killers of old age will not be the things that get you. Scores of other issues in the aging process, both large and small, will have useful solutions available in the broader medical marketplace. We will all live longer and in better health as a result, but no-one will be able to say for just how long until this is tried.

CODA

AND THE LAST WORD IS HIS

The future is not a conveyor belt, but rather something that is built, brick by brick, by the actions of people.

— Fight Aging!, June 2005

Twenty years and one company later, he writes the same argument he opened the book with — except now the field exists, the trials are running, and he is the one running them. The last word is a description, in his own voice, of the medicine he is building.

The Roadmap, 2024

Fight Aging!, 25 March 2024

It has been going on eight years since I last speculated on the order of arrival of the first rejuvenation therapies. Tempus fugit, and time for an updated version! Eight years is a long enough span of time for the first of those rejuvenation therapies to now exist, albeit in a prototypical form, arguably proven in principle but not concretely. The world progresses but my biases remain much the same, rooted in the SENS view of aging set out earlier in this book. The number of groups aiming to produce these therapies has grown considerably. A longevity industry now exists, scores of biotech and pharmaceutical companies where eight years ago there were only a handful. We are eight years further into the grand transition across decades that is taking place in the medical life sciences, in which the treatment of aging will grow to become the majority of medicine.

Age-related diseases are, as covered earlier in this book, simply the names we give to collections of symptoms produced by the same underlying processes of cell and tissue damage that cause aging generally. The full rejuvenation toolkit of the next few decades will consist of a range of different treatments, each targeting a different type of molecular damage in cells and tissues. What follows is a list of potential (and existing!) rejuvenation therapies in a speculative order of arrival.

1. Clearance of Senescent Cells

A good number of companies are presently developing a wide range of senolytic approaches to selectively destroying senescent cells, thereby removing their contribution to degenerative aging. A wide range of small molecules provoke apoptosis in senescent cells in a wide range of different ways, while other approaches train the immune system to better destroy lingering senescent cells. Studies in mice continue to demonstrate rapid, sizable reversal of aspects of aging and many different age-related diseases following the use of senolytic therapies. The most advanced of the first generation senolytic therapies are in phase 2 clinical trials. Given another decade there will be multiple novel senolytic drugs approved for use in the clinic, and no doubt increasingly used off-label by older individuals.

I'll make the argument that the first effective senolytic therapy to be tested in animals, the dasatinib and quercetin combination, is both legitimately a rejuvenation therapy and also readily available to anyone who wants to undergo the treatment. The use of dasatinib and quercetin in combination has been demonstrated to clear senescent cells from some human tissues to a similar degree as it does in mice. It is prescribed off-label by more adventurous physicians and anti-aging practices. It will likely be decades before we know the degree to which any senolytic affects human life expectancy — no-one seems much incentivized to run the sort of long-term human trial that would be required. Nonetheless, the burden of senescent cells is in principle a component of aging, supported by a great deal of animal data. Dasatinib and quercetin removes some of that burden, and this is rejuvenation to my eyes, even if broader benefits to health remain to be demonstrated comprehensively in human trials.

2. Restoration of a Youthful Gut Microbiome

The gut microbiome ages in the sense that the distribution of microbial populations changes in harmful ways: more pro-inflammatory microbes and fewer microbes capable of generating beneficial metabolites. Animal studies suggest that the state of the gut microbiome is at least as consequential to long-term health as diet and exercise. As is the case for senolytics, proven ways to reverse age-related changes in gut microbiome exist, but are not widely used, and it will most likely remain unclear for decades to come as to exactly how much of an effect such a rejuvenation has on human long-term health and life span.

The most obvious and cost-effective intervention for rejuvenation of the aged gut microbiome, the results demonstrated in animal studies and a few small human trials, is fecal microbiota transplant from a young individual. This produces a lasting reset of the gut microbe, and can be readily carried out by anyone willing to put in the work. There are even services that sell screened fecal material from young donors. A second approach is immunization with flagellin to encourage the immune system to clear unwanted microbes, those equipped with flagella. These microbes are largely harmful, causing chronic inflammation, as well as diminishing the population size of beneficial species by outcompeting them.

Unlike senolytics, there is no rush to commercialize forms of fecal microbiota transplant for the treatment of age-related conditions. There is one FDA-approved fecal microbiota transplantation therapy, for a condition in which the intestine is overrun with pathological bacteria, but that is about it. Thus it seems unlikely that concrete data will emerge any time soon on the degree to which gut microbiome rejuvenation improves health and life span. While we can say that it is evidently rejuvenation, and animal data supports that assertion, whether it is rejuvenation to a practical degree in humans remains to be proven.

3. Clearance of the First Few Types of Amyloid

There are about twenty different types of amyloid in the human body, misfolded proteins that form solid deposits. Not all are robustly associated with age-related dysfunction, but of those that are, some progress has been made towards effective therapies based on either direct clearance or interfering in the pace of creation of altered proteins. In the matter of the amyloid-β associated with Alzheimer's disease, there are now several immunotherapies that have demonstrated effective clearance of amyloid-β from the brain. The side-effect profile leaves much to be desired, and it has become clear that late stage Alzheimer's is past the point at which clearing amyloid-β helps all that much. It may well be a useful preventative strategy, however, assuming that the treatments for clearance can be made more benign.

Transthyretin amyloid is associated with heart disease, and is thought to be the primary cause of death in supercentenarians. There are now FDA-approved therapies based on interfering in the creation of altered transthyretin. Some are applicable to the wild-type rather than genetic condition of transthyretin amyloidosis. Arguably every older person should be using these intermittently, assuming a mild side-effect profile, but it will take some time for costs to fall to the point at which this is practical.

This sentiment applies to any therapy targeting forms of amyloid — and there are many more forms to be addressed. To the degree that these treatments are effective and safe, everyone much over the age of 40 should be undergoing a course of treatment every few years.

4. A Robust Cure for Cancer

If asked a decade ago, a universal cure for cancer looked fairly distant. The fundamental logic — block both telomerase and the alternative lengthening of telomeres (ALT) mechanism that cancer depends on to escape the Hayflick limit, as covered earlier in this book — was already understood, but no program looked ready to make the leap to the clinic. Still, all it takes is one innovative approach to produce good enough animal data, and a clinical program will rapidly arise. At present the drug called THIO is in clinical trials after an accelerated program of development at Maia Biotechnology. THIO is metabolized and utilized by telomerase, then incorporated into telomeres to cause disruptive consequences leading to cell death. Since nearly all cancer cells aggressively utilize telomerase, these are the cells that die when THIO is introduced. It should work for the 90% of cancers that do not evolve to make use of ALT, and will be widely used off-label following clinical approval for any one type of cancer. From this starting point, we might expect a great deal more effort in the decades to come to focus on telomere lengthening as a primary target in cancer — and at some point a group with a novel approach will swoop in to deal with the remaining ALT part of the problem.

5. Thymic Rejuvenation to Increase the Supply of Immune Cells

Another possible approach to partially restore lost function in the aging immune system is to increase the pace at which new immune cells are created. This is a very slow pace indeed in older people, due in large part to the age-related decline of the thymus. The thymus acts as a nursery for the maturation of T cells, and its atrophy thus restricts the rate at which new cells enter circulation. Over the 2010s, there was some progress towards engineering of replacement active thymus tissue, as well as methods of providing signal proteins that instruct the old thymus to regenerate and begin to act in a more youthful manner. Transplants of young thymus organs into old mice demonstrated that this class of approach can produce a meaningful improvement in immune function, and thereby extend healthy life.

A decade ago, this was one of a number of regenerative approaches that were on the verge, just waiting for someone to join the final two dots together, found a biotech company, and get moving. That now seems to be happening. The approach of providing signal proteins has proven to be hard, but there are now a few biotech companies, some quite well funded and connected, focused on either (a) cell therapies using cell types that naturally home to the thymus, such as Thymmune Therapeutics, or (b) looking for small molecules that cleverly interfere in the regulation of thymus growth while avoiding the pitfalls associated with past efforts, such as Thymofox.

Further, Intervene Immune has run clinical trials of a growth hormone approach, producing data to suggest a modest degree of thymic regrowth over a year or more of treatment; interestingly data from the CALERIE trial of calorie restriction indicates a similar gain from a few years of slight calorie restriction, implying thymic involution to be perhaps a more dynamic process than suspected. Meanwhile, some interesting advances are taking place in the research community, such as the possibility of a gene therapy delivery system that can in fact effectively target the thymus following intravenous delivery. Exciting times! The state of the field looks promising for some form of effective thymus rejuvenation strategy to emerge in the decade ahead.

6. Mitochondrial Repair

Mitochondrial DNA damage and its consequences for cell function are covered earlier in this book.

There are numerous possible approaches to the problem of dysfunctional mitochondria in aged tissues: upregulation of existing repair mechanisms of mitophagy; delivery of replacement mitochondrial DNA or whole mitochondria; partial reprogramming of cells to restore normal gene expression relating to mitochondrial dynamics and mitophagy; and so forth. Of these, the closest to the clinic are mitochondrial transfusion therapies and the various approaches to adjusting repair and propagation of damaged states in mitochondria, trying to tilt the balance to favor better function. The development of partial reprogramming therapies has enormous funding at present, but arguably much larger challenges must be solved before it can be broadly applied to tissues across the body. It remains hard to say how effective these approaches will be relative to one another. How long will they last before fading in an aged tissue environment? Also unknown.

In the case of mitochondrial transfusion, two companies (Cellvie and Mitrix Bio) are working towards clinical programs. Their challenge is near entirely the development of the processes by which enough mitochondria for human use can be manufactured. From the present starting point that is a tough scaling problem. In the case of altering mitochondrial dynamics or the regulation of mitophagy, there are a range of possibilities already sold in the supplement market, such as SkQ1 and MitoQ, none of which are all that impressive when compared to the effects of exercise. Will companies like Stealth Bio do any better than this? That remains to be seen. There is a market for small molecule and supplement-like products that are not as good as exercise, but it is probably not a market that should interest us.

The SENS approach is somewhat more radical, involving gene therapy to introduce copies of the thirteen genes into the cell nucleus, altered to ensure that the proteins produced can migrate back to the mitochondria where they are needed. Mitochondria will thus have the necessary protein machinery for correct function regardless of the state of their DNA. This has been demonstrated for two of the thirteen genes of interest, numbers two and three in 2016, and getting that far took the better part of ten years at a low level of funding. A company, GenSight Biologics, has championed this approach in clinical trials for one of the genes, in the treatment of a rare genetic disorder, but little further or broader development has taken place outside of academia. Will it be useful to have therapies that fix half the problem, moving six or seven genes to the cell nucleus? Will that reduce the impact on aging by half? It is hard to say until that is possible and demonstrated in mice. A decade ago, it seemed plausible that researchers would get there by now — but this has not happened.

There is still too little funding and support for this approach, and one might well argue that backing mitochondrial transfusion seems a better wager at this point, even given the unknowns.

7. Reversing Stem Cell Aging

The stem cell industry remains massively funded, and is ultimately on a collision course with stem cell aging. Most of the conditions that one would want to use stem cell therapies to treat are age-related conditions. Researchers must thus work towards ensuring that the altered cellular environment, the damage of aging, doesn't prevent these treatments from working — that pristine cells can integrate and work well, not immediately die or decline in response to an age-damaged stem cell niche. Despite some progress over the past decade, particularly around the question of whether cellular senescence is degrading cell therapy outcomes, it is fair to say that the research community isn't engaging aggressively with this goal. Possible reasons for this include the fact that most stem cell treatments, even without addressing issues of the aged tissue environment, represent a considerable improvement in the scope of what is possible to achieve through modern medicine. So the incentive to go further is perhaps not as strong as it might otherwise be.

Stem cell populations become damaged by age, falling into quiescence or declining in overall numbers. They should be replaced with new populations, but while simple in concept, and even achieved for some cell types, such as the blood stem cells that produce immune cells, this is easier said than done for the body as a whole. Every tissue type is its own special case. There are hundreds of types of cell in the body. Each supporting stem cell population has so far required specific methodologies to be developed, and specific behaviors and biochemistry to be laboriously mapped. It isn't even entirely clear that researchers have found all of the stem cell or stem-like cell populations of interest. There is an enormous amount of work to be done here, and at the moment the field is still largely in the phase of getting the basics, the maps, and the reliable therapeutic methods sorted out for a few of the better understood tissue types, bone marrow and muscles in particular. All in all this has the look of a long-term, incremental prospect, despite the high levels of funding for this line of medical research and development.

Are there ways other than complete replacement of cell populations that might enable reinvigoration of aged stem cell populations? It seems that there might be, though we can argue about the degree to which these approaches are in any way affecting aging per se. There are ways to adjust the regulation of stem cell behavior that improve tissue function even given that one is working with aged stem cell populations. To pick one example, Ship of Theseus develops an approach based on Hox family transcription factors that appears to provoke greater muscle stem cell activity. As another example, look at the work of Mogling Bio, building on demonstrations showing CASIN to improve stem cell function in a number of populations, particularly the hematopoietic stem cells of the bone marrow. These and other, analogous approaches will find their way to clinic long before replacement of stem cell populations is a going concern.

8. Clearance of Cross-Links, Glucosepane-Based and Others

Clearance of cross-links — overwhelmingly a matter of the single compound glucosepane, as covered earlier in this book — remains one of the most exciting of early rejuvenation therapies, precisely because it is a single target with such broad influence over vascular stiffness and cardiovascular mortality.

The attention given to glucosepane cross-link clearance remains anemic, despite considerable efforts to create a toolkit and unblock the research community conducted by the SENS Research Foundation and their allies in the research community, including a method of cheaply and reliably synthesizing glucosepane. Work in the mid-2010s conducted in the Spiegel Lab, as well as other parallel lines of research, led to the formation of Revel Pharmaceuticals to commercialize glucosepane cross-link breakers. Unfortunately, little further progress has occurred — while still being a sizable step forward over the state of the field a decade ago, this remains a narrow effort pioneered by few researchers.

A small number of other narrow programs have emerged focused on the lens of the eye, where different forms of cross-linking are involved in stiffening the lens to the point at which muscle strength is inadequate to focus correctly. A lipoic acid choline ester approach looked promising, but failed in phase 2 trials. Another company, Lento Bio, is now in the early stages of working on another approach to the cross-links that stiffen the lens. Again, there are few groups in this space, and more are needed to ensure a good chance of progress towards the clinic in the near future.

9. Partial Reprogramming

Partial reprogramming — the use of brief exposure to Yamanaka factors to reset the epigenetic changes characteristic of aging, covered earlier in this book — cannot help with existing mutational damage to nuclear DNA, nor with the presence of persistent molecules that even youthful cells cannot effectively break down. But it is demonstrated to restore lost mitochondrial function, to pick one example.

When might we expect the first therapies based on partial reprogramming to reach the clinic? This part of the field has become enormously well funded in recent years. Given the $3 billion investment in Altos Labs alone, never mind the other few biotech startup companies with more than $100M in initial funding each, such as Retro Biosciences and NewLimit, there will be no shortage of effort put into preclinical and clinical development. Nonetheless, the challenges are sizable. Different cell types and tissues require different exposures and balances of reprogramming factors for best effect. Too much reprogramming produces cancer via induced pluripotency. Too much reprogramming in the liver and intestines seems very detrimental to health in animal studies. The most likely path to near term therapies is to restrict them to isolated organs, such as the retina and optic nerve. Another possibility is for the groups working on small molecules capable of triggering expression of reprogramming factors to land on something that is more useful for systemic delivery than the gene therapy approaches, but development of small molecule reprogramming is still in the very early stages.

So it remains to be seen as to how matters will unfold in the years ahead. The advent of reprogramming as the leading, widely recognized approach to rejuvenation, and the degree to which funding poured into this project, caught a lot of people by surprise. It seems likely that the surprises will continue, given that we are only a few years into the development of this part of the field.

10. Immune System Destruction and Restoration

The destruction and recreation of the immune system is not a widespread technique, but it has been demonstrated successfully in human clinical trials and animal studies in a variety of contexts over the past twenty years. Researchers and clinicians have used chemotherapy to destroy immune cells and the hematopoietic cells that create them, followed by hematopoietic stem cell transplant (HSCT) to reconstruct the immune system. This approach has resulted in effective cures for multiple sclerosis patients, and has been attempted with varying degrees of success for a number of other autoimmune conditions. The catch here is that chemotherapy and HSCT are not trivial undertakings. The costs and risks are significant, both immediately and in terms of impact on later health and life expectancy. It only makes sense for people who are otherwise on their way to an early death or disability, as is the case for multiple sclerosis patients. However, there are a number of approaches on the way to practical realization that will make chemotherapy obsolete for the selective destruction of immune cells and stem cells — approaches with minimal or no side-effects. Consider a combined approach targeting c-kit and CD47, for example. Sophisticated cell targeting systems such as the gene therapy approach developed for senescent cell clearance by Oisin Biotechnologies could also be turned to stem cell or immune cell destruction, given suitable markers of cell chemistry. There are quite a few of these, any one of which would be good enough.

Replacing the taxing procedures of chemotherapy and HSCT with a safe, side-effect-free treatment would mean that the field of immune system restoration could immediately expand to assess its merits as a treatment for immunosenescence, the age-related failure of the immune system. This decay is in part a problem of configuration: a lifetime of exposure to persistent pathogens such as herpesviruses leaves too much of the immune system uselessly devoted to specific targets that it cannot effectively clear from the body, and too little left ready to fight new threats and destroy malfunctioning cells. Then there are various forms of autoimmunity that become prevalent in older people, not all of which are in any way fully understood — consider the comparatively recent discovery of type 4 diabetes, for example. Clearing out the entire immune system, all of its memory and quirks, and restarting it fresh with a new supply of stem cells is a good approach to many of the issues in the aged immune system. Not all of them, but many of them, and considering the broad influence immune function has over many other aspects of health and tissue function, it seems a worthwhile goal.

That said, has there been much meaningful progress in this part of the field over the past ten years? Not really. Some research has moved forward incrementally, such as on the topic of age-associated B cells and their depletion, or clearance of microglia in the brain. In the broader field, a good number of immunomodulatory approaches to therapy are under development at various stages, some even explicitly aimed at restoring better immune function in the old, but very few target clearance of immune cell populations. The only one that springs to mind is related to the aforementioned clearance of microglia in the brain, since there are existing drugs and a simple mechanism — CSF1R inhibition — by which this can be achieved. The early stage venture Glionics would like to use this clearance and recovery as a way to deliver drugs throughout the brain, rather than specifically to achieve benefits resulting from clearance.

11. Clearance of Other Amyloids, Aggregates, and Sundry Lysosomal Garbage

A good portion of aging — the buildup of lipofuscin, altered tau, and the rest of the junk that our biochemistry cannot break down or clear, as covered earlier in this book — still awaits its dedicated therapies. In many cases there isn't even a good defensible link between a specific waste compound and specific age-related diseases: the waste is one contribution buried in many contributions, and the research community won't start putting numbers to relative importance until it is possible to clear out these contributions one by one and observe the results.

A range of research groups are picking away at individual forms of waste, some with large amounts of funding, some with very little funding, but this is a similar situation to that I outlined above for stem cell aging. There is a huge amount of work to accomplish because there are many targets to address, and with few exceptions, such as amyloid-β, it is unclear which of the targets are the most important. They will all have to be addressed, in some order, but there are only so many researchers and only so much funding. We can hope that as the first effective therapies spread into wider clinical use, treatments for the clearance of forms of amyloid, there will be a growing enthusiasm for work on ways to remove other types of metabolic waste.

What We Are Building Now

Fight Aging!, 26 May 2026

Atherosclerosis — the buildup of fatty plaque that narrows blood vessels until one ruptures to cause a stroke or heart attack, and the mechanism of macrophage overload behind it, covered earlier in this book — remains the most common cause of human mortality.

I founded Repair Biotechnologies with Bill Cherman back in 2018, and the company has since been developing a means to regress atherosclerotic plaque. The biggest challenge for present cardiovascular medicine is that the primary strategy of lowering circulating LDL cholesterol via lifestyle, statins, PCSK9 inhibition, and other more recently developed methods only slows plaque growth, and very slowly, over years, somewhat stabilizes the most unstable plaques. It does not produce sizable or reliable regression of plaque, and this is why atherosclerosis remains the single largest cause of human mortality: 27% of all deaths are via heart attack and stroke, and other varieties of cardiovascular disease caused by obstructive plaque may contribute meaningfully to the demise of another 10–15% of our species. What follows is a recent interview and update on our progress, and an explanation of how our lead drug works.

Developing a Drug to Reverse Heart Disease

So let's get straight into the lead candidate. REP-0003 — can you give us the quick, layman's version on how that works?

Firstly, it's actually REP-0004 now that is the lead candidate. We updated the sequence, but these two drugs are both very similar. They are lipid nanoparticles that encapsulate messenger RNA (mRNA) and then deliver it directly to the liver and nowhere else in the body. It goes to the liver via the normal mechanisms of lipid nanoparticle delivery. The particles are sized to pass through the blood vessel walls for glands like the liver, and the liver is the usual destination for things that are injected intravenously anyway.

Secondly, a liver-specific ligand and the resulting fusion protein target only excess intracellular free cholesterol — unmodified cholesterol, that is — the mechanism of which I've described in detail elsewhere in this book.

This produces a variety of benefits to your liver, because excess free cholesterol serves no useful purpose and is toxic if you have too much of it inside a cell. Normally, a cell will attempt to take that free cholesterol and keep just a little bit of it, and the rest of it gets esterified into lipid droplets or put into the cell membrane or handed off in some way. But, if you get fat or you get old, in both cases, this process stops working as well, and you have too much free cholesterol inside your cells. I should emphasize this really isn't just a problem of obesity. You can have thin people with excessive intracellular free cholesterol in their liver and elsewhere in their bodies.

Reducing free cholesterol in the liver makes the organ behave as if it's in a cholesterol deficit, even though it isn't, so it pulls as much cholesterol back from the rest of the body as it can. Various mechanisms of homeostasis dial up reverse cholesterol transport, dragging free cholesterol from the rest of the body back to the liver, where it then gets intercepted by our fusion protein and broken down. This feedback loop operates for the few days that the protein is present in the body as a result of an mRNA therapy, and the outcome is a draining of excess free cholesterol everywhere, not just the liver. That reduces inflammation. It improves tissue function throughout the body, outside the brain at least, as the brain has its own fairly distinct cholesterol metabolism.

The two outcomes of greatest interest at the moment are, firstly, a dramatic regression of atherosclerotic plaque very rapidly, and secondly, a reversal of liver disease, such as metabolic dysfunction-associated steatohepatitis. You also get a reduction in fibrosis. It happens very rapidly and dramatically, and that is how the drug works.

How receptive has the FDA been to this approach?

They like it. The FDA looked at our materials for the pre-IND and said, "Yes, we agree. Go do this." No, they didn't note any meaningful objections, just the usual small feedback. They granted us an orphan drug designation for treatment of the accelerated atherosclerosis that characterizes the rare disease of homozygous familial hypercholesterolemia, and they don't hand that out like candy. The FDA has to really be interested in what you're doing for that to happen. We were granted orphan drug designation mid last year, and right now we're applying to the Rare Disease Evidence Principles (RDEP) program, which is a new next-level addition to the orphan drug designation. This was announced late last year, and right now nobody really knows, including the FDA, how this program will work out in detail. In principle, it should speed the path to approval for these rare disease therapies where it's somewhat more difficult to organize trials than for a more common condition. The FDA is going to rely more on preclinical evidence and post-approval assessment of the drug than on trials. We'll see how it develops.

Editor's note: This interview was conducted a few months before this book went to press. Repair Biotechnologies has since been granted eligibility for the Rare Disease Evidence Principles program.

About the Author

Reason has written the longevity science blog *Fight Aging!* since 2004, having previously founded and run its predecessor, the Longevity Meme, from 2002. Across more than two decades he has become the field's most persistent chronicler and one of its most consequential advocates, publishing more than twenty thousand posts on the science, politics, funding, and philosophy of treating aging as a medical condition.

He holds graduate degrees in astrophysics and space physics, and worked as a software consultant before turning to biotechnology. He was an early angel investor in longevity startups, including Oisin Biotechnologies, and helped fund the foundational research of the SENS Research Foundation and the Methuselah Foundation.

In 2018 he co-founded Repair Biotechnologies with Bill Cherman, and serves as its chief executive. The company develops therapies to clear the excess cholesterol that drives atherosclerosis; its lead program received Orphan Drug Designation from the United States Food and Drug Administration in 2025.

He continues to write *Fight Aging!*, very nearly every day, at fightaging.org.

Further Reading

The complete archive of *Fight Aging!*, more than twenty thousand posts from 2002 to the present, is freely available at **fightaging.org** under the same Creative Commons license as the writing in this book.

Two longer works, self-published by the author and not reprinted here, extend the material in Part V:

How to Start a Biotech Company in the Longevity Industry (2022) — a practical manual for founders, and the natural sequel to this book's account of the move from advocacy to industry.

A Proposal to Accelerate Progress Towards Human Rejuvenation (2023) — the fuller form of the argument made in "The Off-Label Manifesto."

A Note on Sources

Every chapter in this book was first published as a post at *Fight Aging!* (fightaging.org), under the Creative Commons Attribution 4.0 International License. Each is listed below with its original title's date and web address, so that any reader may find the source, read the surrounding daily record, and confirm what has been selected and what has been cut. Where a chapter began as the author's own transcript of a talk or interview, that too was published at *Fight Aging!* and is licensed on the same terms.

Part I — Aging Is Damage

The Aging Is Damage Thesis — 16 June 2008. https://www.fightaging.org/archives/2008/06/aging

Programmed or Damaged? — 12 June 2015. https://www.fightaging.org/archives/2015/06/yet-more-discussion-of-programmed-aging

Repair, Don't Tinker — 17 December 2013. https://www.fightaging.org/archives/2013/12/extending-life-by-manipulating-metabolism-only-produces-dramatic-results-in-short-lived-species

What SENS Actually Proposes — 15 April 2014. https://www.fightaging.org/archives/2014/04/what-is-robust-mouse-rejuvenation-and-why-should-we-care

Part II — The Damage, One by One

The Rogue Power Plants — 8 October 2006. https://www.fightaging.org/archives/2006/10/how-age-damaged-mitochondria-cause-your-cells-to-damage-you

The Cells That Will Not Die — 21 June 2006. https://www.fightaging.org/archives/2006/06/if-my-cells-liv

The Scandal of Glucosepane — 8 December 2015. https://www.fightaging.org/archives/2015/12/scores-of-labs-should-be-gearing-up-to-work-on-glucosepane-cross-link-breakers-but-are-they

How to Cure Cancer — 24 April 2013. https://www.fightaging.org/archives/2013/04/three-and-a-half-ways-to-cure-cancer

Reprogramming, and What It Isn't — 9 December 2021. https://www.fightaging.org/archives/2021/12/a-review-of-lifespan-the-book-and-some-confusion-about-aging

The Cholesterol That Kills You — 12 August 2020. https://www.fightaging.org/archives/2020/08/reducing-ldl-cholesterol-is-the-wrong-target-for-cardiovascular-disease

The Undefeated Monster — 2 November 2021. https://www.fightaging.org/archives/2021/11/atherosclerosis-the-as-yet-undefeated-monster

Clearing the Plaque — 15 August 2024. https://www.fightaging.org/archives/2024/08/presentation-videos-from-the-2024-rejuvenation-startup-summit-in-berlin

Part III — Why Nobody Listens

Why the Public Shrugs — 26 May 2014. https://www.fightaging.org/archives/2014/05/a-world-of-inadvertently-held-stoic-views

Deathism — 4 March 2007. https://www.fightaging.org/archives/2007/03/what-a-maroon

Death Does Not Give Life Meaning — 16 August 2012. https://www.fightaging.org/archives/2012/08/removing-the-pressure-of-impending-death

The Cost of Living Longer — 2 December 2013. https://www.fightaging.org/archives/2013/12/differences-among-the-naysayers

Is Wanting to Live Selfish? — 11 February 2015. https://www.fightaging.org/archives/2015/02/disappointing-comments-on-longevity-science-from-bill-gates

Gerontocracy and Other Objections — 22 March 2013. https://www.fightaging.org/archives/2013/03/we-already-live-in-a-gerontocracy

Religion and the Long Life — 29 August 2014. https://www.fightaging.org/archives/2014/08/amateur-religious-engineering-and-longevity-science

Mortality and the Shape of Society — 22 October 2012. https://www.fightaging.org/archives/2012/10/breaking-the-wheel-of-time

Better Myths — 17 February 2007. https://www.fightaging.org/archives/2007/02/every-story-is-the-story-of-the-fall

Part IV — Why This Time Is Different

The Fight Over "Anti-Aging" — 11 March 2004. https://www.fightaging.org/archives/2004/03/the-meaning-of-antiaging

Why Now Is Not the 1970s — 21 January 2015. https://www.fightaging.org/archives/2015/01/is-it-different-this-time-around

Nobody Says It's Impossible Anymore — 13 August 2011. https://www.fightaging.org/archives/2011/08/nobody-is-arguing-that-radical-life-extension-is-impossible

It Is 1885 and We Are Arguing About Flight — 15 August 2012. https://www.fightaging.org/archives/2012/08/dont-argue-incrementalism-in-a-time-of-revolutionary-change

Biotech's Cost Curve — 21 January 2014. https://www.fightaging.org/archives/2014/01/the-2010s-in-biotechnology-reflect-the-1960s-in-computing

Against Kurzweil — 21 September 2005. https://www.fightaging.org/archives/2005/09/reading-the-sin

Science Is Not Engineering — 7 January 2015. https://www.fightaging.org/archives/2015/01/the-scientific-institution-is-biased-against-shortcuts-to-the-production-of-practical-technology

The Arcane World — 8 October 2015. https://www.fightaging.org/archives/2015/10/the-arcane

How to Read a Science Story — 16 June 2009. https://www.fightaging.org/archives/2009/06/you-have-to-do-better-than-that

Selling Fasting — 15 October 2016. https://www.fightaging.org/archives/2016/10/a-calorie-restricted-medical-diet-to-be-filed-next-to-selling-ice-to-eskimos

The Debate That Decides How Long You Live — 2 September 2008. https://www.fightaging.org/archives/2008/09/the-scientific-debate-that-will-determine-how-long-we-all-live

Interlude: 2015 — 31 December 2015. https://www.fightaging.org/archives/2015/12/a-look-back-at-2015-in-longevity-science

Part V — Money, Law, and the Machine

A Strategy for Advocacy — 25 May 2011. https://www.fightaging.org/archives/2011/05/three-parallel-tracks

Money Doesn't Matter; Methodology Does — 10 September 2014. https://www.fightaging.org/archives/2014/09/when-google-is-throwing-100-million-at-aging-research-why-fund-sens

Why Rich Sick People Don't Fund Cures — 13 May 2016. https://www.fightaging.org/archives/2016/05/why-do-so-few-wealthy-sick-individuals-fund-medical-research-to-treat-their-conditions

Putting My Own Money In — 17 February 2016. https://www.fightaging.org/archives/2016/02/fight-aging-invests-in-oisin-biotechnologies

The Off-Label Manifesto — 13 January 2023. https://www.fightaging.org/archives/2023/01/a-proposal-to-accelerate-progress-towards-human-rejuvenation

Plausible, Possible, Expensive, Prohibited — 3 May 2012. https://www.fightaging.org/archives/2012/05/plausible-possible-expensive-prohibited

Let Patients Pay — 19 October 2017. https://www.fightaging.org/archives/2017/10/patient-paid-clinical-studies-are-a-good-plan-for-rejuvenation-therapies

Against the FDA — 16 April 2018. https://www.fightaging.org/archives/2018/04/a-nuanced-opposition-to-the-fda-and-similar-regulatory-agencies

Dismantling Efficacy Review — 11 March 2025. https://www.fightaging.org/archives/2025/03/potential-roads-for-upheaval-ahead-in-medical-regulation-in-the-us

Fix Aging First, Politics Second — 26 June 2004. https://www.fightaging.org/archives/2004/06/libertarianism-pragmatism-healthy-life-extension

Death and Taxes — 24 December 2006. https://www.fightaging.org/archives/2006/12/death-by-degenerative-aging-and-taxes

Part VI — Doing It Yourself

What Is Actually Available Today — 29 December 2015. https://www.fightaging.org/archives/2015/12/existing-longevity-therapies-that-are-technically-feasible-but-difficult-and-expensive-to-obtain

Self-Experimentation, Responsibly — 27 April 2020. https://www.fightaging.org/archives/2020/04/how-to-plan-and-carry-out-a-simple-self-experiment-a-single-person-trial-of-flagellin-immunization

The Results, Three Years On — 3 April 2023. https://www.fightaging.org/archives/2023/04/three-years-of-gut-microbiome-data-for-flagellin-immunization-and-fecal-microbiota-transplantation

Rolling Your Own Senolytics — 12 January 2018. https://www.fightaging.org/archives/2018/01/how-to-plan-and-carry-out-a-simple-self-experiment-a-single-person-trial-of-chemotherapeutic-senolytic-drug-candidates

The Underground — 27 July 2009. https://www.fightaging.org/archives/2009/07/the-vegas-group-a-retrospective

Interlude: 2019 — 31 December 2019. https://www.fightaging.org/archives/2019/12/a-look-back-at-2019-progress-towards-the-treatment-of-aging-as-a-medical-condition

Part VII — The Backup Plan

Never Bet on One Horse — 12 October 2015. https://www.fightaging.org/archives/2015/10/cryonics-is-still-the-only-viable-backup-plan

The Moral Urgency — 16 May 2011. https://www.fightaging.org/archives/2011/05/when-did-we-become-suicidal-negligent-barbarians

Why Trusts Won't Save You — 18 June 2009. https://www.fightaging.org/archives/2009/06/cryonics-and-economic-incentives

Why I Signed Up — 16 July 2016. https://www.fightaging.org/archives/2016/07/finally-signed-up-for-cryopreservation-the-existence-of-a-fallback-plan-is-great-but-only-if-you-actually-take-advantage-of-it

An Obituary — 5 December 2014. https://www.fightaging.org/archives/2014/12/leading-from-the-front-to-the-last

Cryonics or Rejuvenation? — 3 March 2022. https://www.fightaging.org/archives/2022/03/should-i-actually-be-working-on-cryonics-rather-than-rejuvenation

Part VIII — What It Is All For

Waking Up — 17 May 2008. https://www.fightaging.org/archives/2008/05/waking-up

The Emergency Room — 21 November 2014. https://www.fightaging.org/archives/2014/11/why-do-we-advocate-for-rejuvenation-research

Another Emergency Room — 8 August 2022. https://www.fightaging.org/archives/2022/08/a-short-commentary-on-why-we-advocate-for-the-treatment-of-aging

On a Death — 1 June 2011. https://www.fightaging.org/archives/2011/06/funerals-are-for-the-living-true-but-more-for-the-ape-inside

A Copy of You Is Not You — 4 September 2012. https://www.fightaging.org/archives/2012/09/personal-survival-and-swimming-against-the-cultural-currents

Immortality, Seriously — 23 January 2017. https://www.fightaging.org/archives/2017/01/the-million-year-life-span

The Ethics of Paradise Engineering — 10 July 2018. https://www.fightaging.org/archives/2018/07/the-hedonistic-imperative-followed-to-the-ends-of-paradise-engineering

Why I Do This — 5 August 2019. https://www.fightaging.org/archives/2019/08/an-interview-with-reason-at-undoing-aging-2019

Escape Velocity — 18 November 2015. https://www.fightaging.org/archives/2015/11/the-immortals-among-us

The Last Mortals — 22 May 2019. https://www.fightaging.org/archives/2019/05/considering-the-experience-of-being-one-of-the-last-mortals

A Letter From the Future — 29 May 2018. https://www.fightaging.org/archives/2018/05/when-i-am-eighty-five

Coda — and the last word is his

The Roadmap, 2024 — 25 March 2024. https://www.fightaging.org/archives/2024/03/predicting-the-order-of-arrival-of-the-first-rejuvenation-therapies-2

What We Are Building Now — 26 May 2026. https://www.fightaging.org/archives/2026/05/an-update-on-progress-at-repair-biotechnologies-developing-means-to-regress-atherosclerotic-plaque

www.ingramcontent.com/pod-product-compliance
Lightning Source LLC
LaVergne TN
LVHW052226200926
848481LV00001B/223
9781951974121